Conflict and I

CW00551726

There is increasing policy and academic awareness of the considerable relationship between conflicts and development. Developmental
factors and inequalities can trigger violence, but can also be imperative in ending conflicts and helping societies with reconstruction and
reconciliation. This second edition has been fully updated and
explores the complexity of the links between violent conflict (including civil wars) and development, underdevelopment, de-development
and uneven development. By emphasizing the connections between
stable developed economies and civil wars in other parts of the world,
Conflict and Development examines how structural factors (such as
the organization of the global economy) can be seen to have virtually
condemned some regions to conflict and underdevelopment.

An invaluable introductory text, the book attempts to respond to a number of critical, persisting and challenging concerns. The most important
of these is power: who holds it, how did they acquire it and can it be
taken from them? Mindful of the power that academics and policymakers have to overwrite the experiences of those in conflict and development contexts, Mac Ginty and Williams strive to consider these issues
from the 'local' perspective, focusing on the on-the-ground experiences
of communities – populations that often see peacebuilding and development as something that is done to them. This book attempts to 'write
people in', bringing the human element, and indeed the human cost,
into explanations of conflict, peace and development.

Accessible, interesting and policy relevant, *Conflict and Development*
is the essential resource for undergraduates looking to better understand the difficult and almost always complex relationships between
armed conflict and development. Drawing on contemporary theoretical
debates and examining current policies and recent events, the book
considers how sustainable existing governmental and NGO guidelines
are for peacemaking, peacebuilding and post-war reconstruction

efforts. The text is illuminated throughout with case studies drawn from Africa, the Balkans, Asia and the Middle East.

Roger Mac Ginty is Professor of Peace and Conflict Studies at the University of Manchester. He edits the journal *Peacebuilding*, and is editor of *The Routledge Handbook of Peacebuilding* and (with Jenny Peterson) *The Routledge Companion to Humanitarian Action*. He has conducted extensive field research in conflict-affected countries, and his latest monograph is *International Peacebuilding and Local Resistance: Hybrid forms of peace*.

Andrew Williams is Professor of International Relations, University of St Andrews. He specializes in the study of conflict and international history. His main research interests include international conflict resolution, international history and international organization. He has had a great deal of experience of the practice of conflict resolution and has worked as a consultant for United Nations organizations including UNDP and UNITAR. He has published widely in key journals. His book *Liberalism and War* was published by Routledge in 2006.

Routledge Perspectives on Development

Series Editor: Professor Tony Binns, University of Otago

Since it was established in 2000, the same year as the Millennium Development Goals were set by the United Nations, the *Routledge Perspectives on Development* series has become the pre-eminent international textbook series on key development issues. Written by leading authors in their fields, the books have been popular with academics and students working in disciplines such as anthropology, economics, geography, international relations, politics and sociology. The series has also proved to be of particular interest to those working in interdisciplinary fields, such as area studies (African, Asian and Latin American studies), development studies, environmental studies, peace and conflict studies, rural and urban studies, travel and tourism.

If you would like to submit a book proposal for the series, please contact the Series Editor, Tony Binns, on: jab@geography.otago.ac.nz.

Published:

Third World Cities, 2nd Edition
David W. Drakakis-Smith

Rural–Urban Interaction in the Developing World
Kenny Lynch

Environmental Management and Development
Chris Barrow

Southeast Asian Development
Andrew McGregor

Postcolonialism and Development
Cheryl McEwan

Disaster and Development
Andrew E. Collins

Non-Governmental Organizations and Development
David Lewis and Nazneen Kanji

Gender and Development, 2nd Edition
Janet Momsen

Economics and Development Studies
Michael Tribe, Frederick Nixson and Andy Sumner

Water Resources and Development
Clive Agnew and Philip Woodhouse

Theories and Practices of
Development, 2nd Edition
Katie Willis

Food and Development
E. M. Young

An Introduction to Sustainable
Development, 4th Edition
Jennifer Elliott

Latin American Development
Julie Cupples

Religions and Development
Emma Tomalin

Development Organizations
Rebecca Schaaf

Climate Change and
Development
*Thomas Tanner and Leo
Horn-Phathanothai*

Global Finance and Development
David Hudson

Population and Development,
2nd Edition
W. T. S. Gould

Conservation and Development
*Andrew Newsham and Shonil
Bhagwat*

Tourism and Development in the
Developing World, 2nd Edition
*David J. Telfer and Richard
Sharpley*

Conflict and Development,
2nd Edition
*Roger Mac Ginty and Andrew
Williams*

Cities and Development,
2nd Edition
Sean Fox and Tom Goodfellow

Forthcoming:

Children, Youth and
Development, 2nd Edition
Nicola Ansell

South Asian Development
Trevor Birkenholtz

Gender and Development,
3rd Edition
Janet Momsen

Natural Resource Extraction
and Development
*Roy Maconachie and
Gavin M. Hilson*

Conflict and Development

Second Edition

Roger Mac Ginty and Andrew Williams

Routledge
Taylor & Francis Group

LONDON AND NEW YORK

Second edition published 2016
by Routledge
2 Park Square, Milton Park, Abingdon, Oxon OX14 4RN

and by Routledge
711 Third Avenue, New York, NY 10017

Routledge is an imprint of the Taylor & Francis Group, an informa business

First edition published by Routledge 2009

British Library Cataloguing in Publication Data
A catalogue record for this book is available from the British Library

Library of Congress Cataloging in Publication Data
Names: Mac Ginty, Roger, 1970-
Williams, Andrew, 1951-
Title: Conflict and development / Roger Mac Ginty and Andrew Williams.
Description: Second Edition.
New York : Routledge, 2016.
Series: Routledge perspectives on development series
Revised edition of the authors' Conflict and development, 2009.
Includes bibliographical references and index.
Identifiers: LCCN 2015032845
Subjects: LCSH: Civil war–Economic aspects.
Economic development–Political aspects.
Classification: LCC HB195 .M195 2016
DDC 338.9–dc23
LC record available at http://lccn.loc.gov/2015032845

ISBN: 978-1-138-88750-3 (hbk)
ISBN: 978-1-138-88752-7 (pbk)
ISBN: 978-1-315-71405-9 (ebk)

Typeset in Times New Roman and Franklin Gothic
by Cenveo Publisher Services

This book is dedicated to Charlie Williams, Private, RAMC, 1896–1916, a working man who lost his life in a war over power and resources, like so many others before and since.

Contents

Plates

Boxes

Abbreviations

ADB	Asian Development Bank
AFDB	African Development Bank
AIIB	Asian Infrastructure Investment Bank
ANC	African National Congress
ATSG	Aid, Security, Trade, Governance
AU	African Union
BiH	Bosnia-Herzegovina
BRICS	Brazil, Russia, India, China, South Africa
CNN	Cable Network News
COIN	counter-insurgency
CPA	Coalition Provisional Authority
DAC	Development Assistance Countries
DDR	disarmament, demobilization and reintegration
DFID	Department for International Development
DRC	Democratic Republic of Congo
EAR	European Agency for Reconstruction
EBRD	European Bank for Reconstruction and Development
ECHO	European Commission Humanitarian Aid Office
ECOWAS	Economic Community of West African States
EU	European Union
FTSE	Financial Times Stock Exchange
GDP	Gross Domestic Product

GNI	Gross National Income
GNP	Gross National Product
ICC	International Criminal Court
ICRC	International Committee of the Red Cross
IFI	international financial institution
IGO	intergovernmental organization
IMF	International Monetary Fund
INGO	international non-governmental organization
INSTRAW	(UN) International Research and Training Institute for the Advancement of Women
IS	Islamic State
ISIL	Islamic State of Iraq and the Levant
LDCs	least developed countries
LTTE	Liberation Tigers of Tamil Eelam
MDGs	Millennium Development Goals
MSF	Médecins Sans Frontières
NATO	North Atlantic Treaty Organization
NGO	non-governmental organization
NICs	newly industrialized countries
OCHA	Office for the Coordination of Humanitarian Affairs
ODA	Official Development Assistance
OECD	Organization for Economic Cooperation and Development
OSCE	Organization for Security and Cooperation in Europe
PA	Palestinian Authority
PIOOM	Interdisciplinary Research Programme on Root Causes of Human Rights Violations
PLO	Palestine Liberation Organization
PMC	private military contractor
PRT	provincial reconstruction team
R2P	responsibility to protect
RUF	Revolutionary United Front (Sierra Leone)
SAP	structural adjustment programme
SDGs	Sustainable Development Goals
SSR	security sector reform
TRC	truth and reconciliation commission
UN	United Nations
UNAVEM	United Nations Angola Verification Missions

UNCTAD	United Nations Conference on Trade and Development
UNDP	United Nations Development Programme
UNICEF	United Nations Children's Fund
UNIDIR	United Nations Institute for Disarmament Research
UNMIK	United Nations Mission in Kosovo
UNRRA	United Nations Relief and Rehabilitation Agency
UNTAC	United Nations Transitional Authority in Cambodia
UNTAET	United Nations Transitional Administration in East Timor
USAID	United States Agency for International Development
WCT	war crimes tribunal

Introduction

Welcome! This book seeks to explain the links between conflict and development in a way that is both accessible and relevant. The book covers the main actors, theories and trends that shape how development and armed conflict impact on one another. It also seeks to explain how we – as scholars, citizens, voters, consumers and investors – are implicated in the complex relationship between conflict and development.

If taken individually, development and conflict raise near-insoluble questions.

Development

- What constitutes development? Is it just about economic issues or does it also extend to quality of life, culture and politics?
- How can we accommodate a rising world population (from 7.2 billion people today to 9.6 billion in 2050) with protection of scarce resources?
- What are the best ways to address fairness and inequality? Should we even bother?
- How can we measure development?

Conflict

- Can we pre-empt armed conflict by identifying the triggers and causes?
- Why do the costs of conflict fall more heavily on some groups?
- What can we do about persistent conflicts, like in Israel–Palestine? Are they simply destined to be with us for ever?
- What are the best ways for collectives of countries (like the United Nations or a coalition) to deal with conflict?

When we bring conflict and development together we also face tough questions. But by combining the two we can see how it is difficult to address conflict without also addressing development. It also becomes clear that development issues can be the key to escalating and de-escalating conflict. The links between conflict and development are seemingly endless. Countries that manufacture arms may want to maximize exports, but these exports may fuel conflict and oppression overseas. South East Asia is becoming increasingly militarized as states make claims on the South China Sea and the mineral wealth beneath it. In recent times the United States has become embroiled in costly, and ultimately unwinnable, wars in Afghanistan and Iraq. But it has only been able to do so by taking out enormous loans. So modern war depends on credit. Non-state armed groups like Islamic State or the Taliban need money to pay fighters and buy arms. They need to exploit markets to raise these funds.

This book seeks to unpack the links between conflict and development. It is mindful though that the picture is very complicated and that we, as observers from the global North, may not always have the best vantage point from which to see conflict and development in other parts of the world. We should be aware of the limitations of predominantly Western research methodologies and methods of explanation. We need to be aware that many of our discussions of conflict and development use peculiarly Western tools of analysis and occur at a level of abstraction far removed from the lived experience of those facing the challenges of conflict and development. Vibrant and diverse communities are rendered into figures in a spreadsheet or are described in the dry tones of economists and conflict analysts. Most development and peacebuilding policy (and the research that underpins it) is designed in the global North but is directed at the global South (Scholey 2006: 179–80; Cohen 2014). Moreover, Western analyses of developing world and conflict contexts can be prone to stereotypes that overwrite a more complex reality. Thus we are all familiar with African women portrayed as '[a]lways poor, powerless and invariably pregnant' or African men as aggressive (Win 2007: 79).

Conflict and development – often studied separately

One of the oddities of social science research has been that theories of development and theories of conflict have largely evolved in isolation from one another. This was especially odd since development economics as an academic subdiscipline and area of policy expertise emerged in a post-Second World War context defined by violent conflict. Moreover, the lens of development studies was firmly focused on the developing states, many of which were prone to conflicts relating to decolonization, post-independence power struggles and proxy competition among Cold Warriors. The few development economists who considered the matter saw war as an interruption of development and surmised that development could not begin until war had ceased in a particular location (Thomas 2006: 186). As a result, most development specialists excluded countries experiencing violent conflict from their studies. In effect, conflict was written out of development.

We must not be too harsh on development theorists: many were economists more at ease with the study of a state's fiscal levers than its political machinations. Moreover, the vast majority of development research was country specific and thus poorly placed to observe regional or international patterns, including those that contributed to violent conflict. Just as development theorists largely ignored conflict, the emerging subdiscipline of conflict studies had little to say – until relatively recently – about development issues. There were a few honourable exceptions among the conflict theorists (Gurr 1970; Azar 1990), but by and large their lenses of inquiry overlooked the potential of development (and de-development, underdevelopment and uneven development) to contribute to both war and peace. The sole interest of many theorists was inter-state war and the interplay between military and political leaders in different states. Economic conditions within states were important only in terms of how they could sustain a state's ability to pursue war. The condition of people within states, and how this might contribute to conflict and popular radicalization, was often overlooked.

The not so splendid isolation of development and conflict studies from one another was no longer sustainable when the end of the Cold War witnessed an upsurge in civil wars. Not only were these post-1991 civil wars more visible than their Cold War predecessors, but the international system was also in flux, with national, regional and international mechanisms struggling to cope. In this context, there was a rush to explain the phenomenon of civil war, and the corpus of academic, policy-related and journalistic work exploring the linkages between conflict and development began to grow. Crucial in this process was a more sophisticated understanding of the nature of conflict and the long-term and often unsatisfactory nature of pacification strategies

(Kaldor 2007). The orthodoxy that development was an integral part of peacebuilding spread very quickly and was notably promoted by UN Secretary General Boutros-Ghali's 1992 *Agenda for Peace* document. There was also a growing realization of the links between the outbreak of violent conflict and underdevelopment and uneven development (Collier 2000b; International Development Committee 2006: 8). As a result, many academics, policymakers and political leaders came to see development as a key to conflict prevention. The concept of human security, which gained prominence from the mid-1990s onwards, was highly influential in broadening conceptualizations of security to encompass issues traditionally regarded as germane to development or social improvement (United Nations Development Programme (UNDP) 1994; Shaw et al. 2006: 3–18).

About this book

Controversies abound in the study of conflict and development: are internal or external factors primarily responsible for a state's underdevelopment? Should aid agencies cooperate with warlords to distribute humanitarian goods? How can the security demands made by Western states be reconciled with human rights and political liberalization in developing world states? What are the implications for humanitarianism and the principle of impartiality of growing civil–military cooperation? Can increasingly intrusive Western means of development programming and good governance be reconciled with indigenous and traditional norms? How can increasingly technocratic development and peace-support interventions address the affective or emotional dimensions of development and peacebuilding? How can powerful corporations be held to account? Can technology solve some long-term development problems? Do we, as researchers, have the correct tools with which to understand conflict and development issues? These questions, and many more, form the basis for this book.

The essential purpose of the book is to chart our understanding of the complex relationship between conflict and development. The book is written from a largely critical perspective: critical in the sense that it questions orthodox assumptions and conventional wisdom, but hopefully the book does not forget that academic sophistry is of little help to those facing the very real hardships that attend the problems of conflict and development. At heart, development and conflict revolve around people, yet social scientists (along with government planners, non-governmental organization (NGO) log-frames and financial models) have been particularly successful in writing people out of their analyses. This book recognizes that humanity red and raw plays an essential part in stories of conflict and development. Hate, rage, revenge, hopelessness, bitterness, ignorance, love, joy and mercy are as relevant to analyses of processes of conflict and development as academic conceptualizations.

Assumptions

The book is based on five assumptions about development that help with our understanding of the connections between conflict and development:

- Development is not necessarily a good thing; it can have negative and unintended consequences.
- Development can trigger and sustain violent conflict.
- Development is an uneven process.
- Development is not just about economic growth.
- Development can be targeted in ways that aid post-war reconstruction and reconciliation.

Cutting across all of these assumptions is the issue of power: who holds it, how did they get it, and can it be taken off them (paraphrased from Benn 2001)? Power comes in many forms – political, economic, official, unofficial, hard (military), soft (diplomatic and cultural), earned, elected, seized and coveted. Power and legitimacy do not always overlap: governments, big businesses and landowners may have official or coercive power, but they may be popularly despised. Local leaders (for example, clerics or tribal leaders) may have little economic or formal political power, but they may be respected within their community and have legitimacy. External actors such as international organizations, donor governments or multinational corporations may have immense power and capability, but again this does not always translate into local acceptance.

When discussing the five assumptions that underpin this book, it is worth considering how they are modified by power. For example, in relation to the third assumption (development is an uneven process) it is worth thinking about what power factors make and maintain the unevenness of development. Do these factors derive from the internal dynamics of the country, or do they relate to international structures?

The first assumption (that development is not necessarily a good thing) may initially seem Luddite or somehow antithetical to human advancement, especially since development is regarded by many as the means through which public goods (education and health care) and personal liberty (freedom of expression and action) can be attained. 'In everyday usage "development" is virtually synonymous with "progress"' (Thomas 2006: 187). Moreover, in the Western political mind, the continuation of economic development is a fundamental assumption of political and economic life. The shelf life of the Western political leader who advocated limits on growth would be very short indeed. Jimmy Carter's emphasis on American fuel dependency is believed to have been a major factor in his 1980 presidential electoral defeat to Ronald Reagan (Homans 2012). Indeed Reagan's 'It's morning in America again' television commercial for his 1984 re-election campaign is regarded as a modern masterpiece in light and fluffy electioneering that shied away from pressing issues such as accelerating economic disparities (Troy 2007). Despite the tremendous political, economic and moral power behind the orthodox position that development is always a 'good thing', critical observers must be prepared to judge development according to its actual impact and ambition. Such judgements will depend on the moral–ethical–political framework held by the individual, community or institution that makes the judgement, and the vantage point from which they make their judgement. Put simply, where one sits will determine how one judges development.

If development is judged against the simple criteria of jobs created or the amount of electricity created (in the case of a hydroelectric dam), then it might be relatively easy to judge a scheme a success or failure. Yet a broader assessment framework is required – one that takes into account unanticipated outcomes and difficult-to-measure consequences such as the loss of culture and scenery. If development is demonstrated to inflame conflict, degrade environmental conditions and have profoundly negative social and cultural consequences, then it is entirely reasonable that observers reflect this in their judgements. It is unreasonable to brand those who are against 'bad' or unjust development as being against all development.

This leads to the second assumption, that development can trigger and sustain violent conflict. This is by no means always the case. Indeed, as will be demonstrated in later chapters, development can help prevent conflict and aid post-conflict reconciliation. Yet is it important to recognize the conflict-promoting potential of development processes whereby inter-group resource competition, population displacement, environmental degradation and the erosion of social structures that may have once restrained conflict may all contribute to violent conflict. These issues will be explored in detail in Chapter 1.

The third assumption, that development, by its very nature, is an uneven process, contributes much to explanations of the initiation, maintenance and ending of violent conflicts. A complex array of structural and proximate economic, political and geographical factors accounts for the uneven nature of development. Many of these factors will be discussed in later chapters. The unevenness of development (and resources and approaches to development) suggests that conflict will be 'inevitable', but as we will see, it is often the management of resources and development that matters. As important as unevenness is the *perception* of unevenness. In societies with identity-based divisions, groups may be anxious to get their 'fair' share of resources. Many citizens in Lebanon, for example, see charities ministering to Syrian refugees and believe that Syrians get preferential treatment when it comes to health care (Harb and Saab 2014: 8)

The fourth assumption, that development is not just about economic growth, stems from the tendency of many observers (not just development economists) to overlook the social, political and cultural dimensions of development. Crucial in this regard is the type of development strategy pursued and the relative importance attached to redistribution and market freedom, as well as political and cultural development. A narrow economic lens could examine China's astounding economic growth rates (averaging at just over 10 per cent between 2005 and 2013) and declare it a development success (World Bank 2013). But a more holistic approach might take account of the state's poor human rights record. Amnesty International's (2013) *State of the World's Human Rights* report observed that

> The authorities maintained a stranglehold on political activists, human rights defenders and online activists, subjecting many to harassment, intimidation, arbitrary detention and enforced disappearance ... Access to justice remained elusive for many, resulting in millions of people petitioning the government to complain of injustices and seek redress outside the formal legal system. Muslims, Buddhists and Christians, who practised their religion outside officially sanctioned channels, and Falun Gong practitioners, were tortured, harassed, arbitrarily detained,

imprisoned and faced other serious restrictions on their right to freedom of religion. Local governments continued to rely on land sales to
fund stimulus projects that resulted in the forced eviction of thousands
of people from their homes or land throughout the country. The
authorities reported that they would further tighten the judicial process
in death penalty cases; however thousands were executed.

(Amnesty International 2013)

The final assumption is to recognize the potential of development to
contribute to post-war reconstruction and reconciliation. Development
can, in the correct circumstances, ease or cement a route out of conflict
and division. In an ideal situation a mutually reinforcing relationship
can be established between development and reconstruction. Formerly
divided peoples can come together for the joint pursuit of economic
growth and social progress. This is not always the case, and poorly
managed post-war reconstruction often has profound consequences for
the nature of the post-war society, some of them negative.

Alongside these assumptions on development, it is worth noting that
the main focus of this book is on civil war, rather than inter-state war.
This is primarily because war between states is a rare phenomenon,
while civil war (albeit often internationalized) is more common. Most
civil wars spill over national boundaries, with arms, refugees, fighters
and resources crossing borders to give conflicts a messy transnational
dimension. In 2014, for example, although 40 armed conflicts were
ongoing, only one of them (between India and Pakistan) could be
classed as inter-state (Pettersson and Wallensteen 2015: 537) Importantly our interest in conflict extends beyond the direct violence of
overt war. Indirect or structural violence plays a key role in contexts
of conflict and development. Such violence is often embedded in the
socio-economic and politico-cultural behaviour of a society. It can be
insidious, barely visible and taken for granted. It takes the form of
discrimination in the provision of public goods and opportunities, the
militarization of society and the prevalence of societal attitudes
(sometimes encouraged by the state or other institutions) that certain
groups are inferior to others. So this book proceeds by adopting
holistic views of both conflict and development. Neither concept
constitutes a neatly compartmentalized category. Instead they are
messy, ill-defined, and there is little agreement about the best way to
pursue development and conflict transformation. Above all, both
concepts relate to the most contrary and awkward species of all:
humans. People don't always behave as we expect them to; they can

Plate 1 *A child's shoe in rubble in Beirut: one of the aims of this book is to write people back into accounts of conflict and development.*

act in ways that strike us as 'irrational'; and they're not always as grateful to us as we feel they should be. People are often written out of analyses of conflict and development.

An array of factors helps write people out of many studies of peace, conflict and development: the media's need to compress thousands of individual experiences into a single narrative; the technocratic bias of policymaking in which units and spreadsheets are more manageable than people; and social sciences' move towards large-scale studies and their inability to deal with the affective dimension of human behaviour. Indicative of this 'writing out' of people is development and conflict-sensitive 'programming' as practised by many donor governments and agencies. The term 'programming' suggests a machine-like process whereby carefully regulated inputs are expected to have particular outputs. Just as war is often criticized for dehumanizing individuals and objectifying them into a lumpen enemy, there is a real danger that responses to conflict and underdevelopment have a similar effect. Most studies, apart from biography and some forms of anthropology, cannot hope to convey individual experiences. It is important that we recognize that people in conflict zones do not constitute an undifferentiated mass and that individuals often experience events and processes in very different ways.

This chapter proceeds with brief overviews of the evolution of theories of development and conflict so as to provide a context for subsequent chapters. In relation to development studies, it is particularly important to understand the contemporary dominance of neo-liberalism and market-led 'solutions' and the consequences of this for internationally-sponsored development and peace-support interventions. The evolution of conflict studies shows how a more complex understanding of the causes and maintenance of conflict has emerged, and how this has shaped contemporary conflict transformation interventions. The chapter concludes by outlining the structure of the book.

The evolution of development theory

Summarizing a multidisciplinary endeavour such as development studies is a difficult task, especially when the ultimate purpose of development, and the optimum means of achieving it, are hotly contested. Kothari (2005: 1) is rightfully critical of surveys of the discipline that 'articulate a singular theoretical genealogy'. Development studies has been peculiarly faddish, seizing upon theories and techniques at particular moments, only to discard them in favour of a new saviour theory or technique. Kothari (2005: 2) also warns against interpretations of development studies that regard 1945 as Year Zero, as though no development or thinking about development occurred before that year. Scrutiny of the means to achieve economic development has a long intellectual pedigree (Meier and Rauch 2000). J. M. Keynes was particularly prescient of the need for peace negotiations to take seriously the issue of long-term economic development (Davenport-Hines 2015). Dispirited after his ringside seat at the Treaty of Versailles negotiations following the First World War, he observed that 'Peace has been declared at Paris. But winter approaches' (Keynes 1920: 235):

> The Treaty includes no provisions for the economic rehabilitation of Europe, – nothing to make the defeated Central Empires into good neighbours ... nothing to reclaim Russia; nor does it promote in any way a compact of economic solidarity amongst the Allies themselves; no arrangement was reached at Paris for restoring the disordered finances of France and Italy, or to adjust the systems of the Old World and the New.
> (Keynes 1920: 235)

As will become clear in later chapters, lessons seem to have been learned by some, in that it is well recognized that the ending of wars provides unique opportunities for international political and economic intervention. Less clear, however, are the optimal types and extent of any intervention.

Although 1945 was not Year Zero, it was the year in which the contours of the modern international financial architecture were established. The World Bank and the Bretton Woods exchange rate mechanism (which facilitates international trade) date from this era. The Second World War also saw the United States re-establish itself as the predominant state in the world economy (Endres 2012). The decade and a half after the Second World War was the highpoint of economic planning during which development economists were convinced that "'good" scientific analysis would generate the "right answers"' (Harriss 2005a: 19). This was the period of positivist orthodoxy, in which the primary aim of development was accelerated economic growth, the primary agent was the state (mediated by the Bretton Woods institutions) and the primary means of achieving growth was careful analysis followed by a precise plan. The plan often involved raising rural productivity and transferring underutilized labour from the agricultural to industrial sectors (Leys 1996: 8). It was an era of optimism, but was also tinged with colonial attitudes: if only those underdeveloped people would adopt scientifically proven methods of agriculture and economic organization then they will become developed – just like 'us'.

By the late 1950s and early 1960s, it was becoming clear that economic growth was difficult to achieve in many developing world contexts, and that the fruits of any growth were rarely shared fairly. There was no shortage of economic theories, but as J. K. Galbraith (1964: 38) observed, 'it would be a mistake to identify complexity with completeness and sophistication with wisdom.' In other words, neat theories did not always work in real world circumstances. The failure of the initial post-war development planning led to a new emphasis on modernization and technology transfer. The key here was the adoption of 'modern' forms of administrative and political organization in imitation of Western states and businesses, the transfer of Western knowledge and skills through education programmes, and a 'green revolution' of increased agricultural productivity through the use of Western farming methods (Rostow 1960). This, in turn, sparked a radical critique in the late 1960s and early 1970s as it became clear that, in many cases, modernization strategies had made few appreciable differences to citizens in the developing world (Harriss 2005a: 19–25). According to left-wing critics, the '"modernising elites" were really ... lumpen-bourgeoisies, serving their own and foreign interests, not those of the people; world trade perpetuated structures of underdevelopment' (Leys 1996: 12). According to this view (usually called 'dependency theory'), modernization was a recipe for further immiseration and a structural dependence on Western economic powers (Frank 1967).

Global political trends, and specifically the Cold War, had a profound impact on development strategies. A number of states, including Cuba, Ethiopia, Tanzania and Vietnam, adopted socialist development programmes, usually under the tutelage and protection of the Soviet Union. To differing degrees, 'scientific socialism' was mobilized in the service of development. This often involved the nationalization of industry, restrictions on foreign capital, land reform, and an enhanced role for the state in directing economic exchanges and initiatives (Clapham 1987). In most cases, the results were not good: the socialist 'reforms' tended to be disruptive and often reliant on state coercion and the global economy offered a poor fit for those not willing to reform. It is worth noting that the socialist experiment states such as Angola, Mozambique and Ethiopia were often the scenes of violent civil war; wars that were often stoked by the US and its allies. The true fragility of these states' economic models did not become apparent until the collapse of the Soviet Union in 1991 and the withdrawal of subventions. Both dependency theory and the socialist model of development paled into the background with the coming tide of neo-liberalism.

Neo-liberalism is the belief that the market, freed from regulation, offered the solution to development problems. The neo-liberal revolution of the 1980s and beyond was in sympathy with wider political and economic changes. The political right was on the ascendant in the United Kingdom and the United States, and globalizing market forces meant that national and international controls over capital had been severely eroded. Traditional responses of state-directed development policy were no longer effective in an economic climate characterized by economic shocks, unstable commodity prices, international capital flight and increasingly powerful and mobile multinational corporations. The end of the Cold War meant not only the collapse of the Soviet Union as a material supporter of alternative models of economic development, but also the collapse of the notion of 'an alternative'.

There was no Plan B any more. Former Soviet satellite states proved to be in no position to resist aggressive economic reform interventions (often called the 'shock doctrine') by international financial institutions. The rise of neo-liberalism was also assisted by abundant evidence of state incompetence and corruption throughout the developing world. Deepak Lal (1998: 65) noted how the 'old development economics ... implicitly assumed that the state was benevolent, omniscient, and omnipotent.' The 'Chicago School' (who claim intellectual authorship of neo-liberalism) were pushing at an open door at the headquarters of the World Bank, the International Monetary Fund (IMF) and right-wing governments. The essential neo-liberal argument was that the dead hand of the state and an invariably bloated public sector acted as a brake on economic development, while an unfettered market could act as an engine of development. In this view, the benefits of market-driven growth would trickle down and benefit all.

According to James Dorn (1998: 13), 'the real plight of underdeveloped countries is not market failure but government failure – that is, the failure of government to protect property rights, enforce contracts, and leave the market alone.' Champions of neo-liberalism were forthright in what needed to be done: the state must be pared back, the market freed from regulation, state assets should be privatized, and exchange controls and industrial licences should be lifted. The salvation for underperforming economies lay in more exposure to the market, not protection from it. Peter Bauer (1998: 36) had little time for 'unfounded notions about Western responsibility for Third World backwardness'. For him it was

> abundantly evident throughout the Third World [that] the poorest and most backward societies and areas are those which have fewest commercial contacts with the West, and the most advanced are those with the most extensive and diversified contacts, including contacts with those bogeymen, the Western multinationals. Throughout the Third World the level of economic attainment declines as one moves away from regions with most Western contacts to the aborigines and pygmies at the other end of the spectrum.
>
> (Bauer 1998: 28)

Neo-liberal prescriptions, which were enthusiastically endorsed by global capital and the leading international financial institutions, were rolled out in former Soviet bloc states, often with catastrophic social consequences (Klein 2007: 180–4). Western governments, and by extension their development aid institutions, increasingly adopted market-led 'solutions' as the best way to assist societies in development and reconstruction.

Cloaked in a populist mantle of the empowerment of entrepreneurs and the cutting of public sector waste, neo-liberal truisms became the new orthodoxy. According to David Harvey,

> Neo-liberalism has, in short, become hegemonic as a mode of discourse. It has pervasive effects on ways of thought to the point where it has become incorporated into the common-sense way many of us interpret, live in, and understand the world.
>
> (Harvey 2005: 3)

In many Western European states we have seen the dismantling of the post-Second World War social contract. State provision, particularly on welfare, has been pared back, and with it, the expectation that the state should provide. The intellectual dominance of neo-liberalism is essential to our understanding of the responses of leading states, international organizations and international financial institutions to the problems of conflict, development and peacebuilding. Market-led 'solutions' have been hardwired into the organizational culture and policy responses of donor governments and NGOs. Whether this manifests itself in micro-credit schemes to further women's empowerment, or in the privatization of state resources as part of a post-war reconstruction programme, it has had a profound impact on the ethos of humanitarianism, development and conflict amelioration policies. Somewhat ironically, the financial crisis that began in 2007, caused by a lack of control on lending to governments and individuals, reinforced neo-liberal 'solutions'. The language of austerity has been used to further cut back the state and normalize the notion that the market is the best, and most sustainable, route to economic salvation.

Under the guise of 'good governance', massive programmes of social, economic and political engineering have taken place in societies emerging from conflict and economic crisis. In many cases, relationships between citizens and the state, the state and the market, and the state and other states have changed radically. For example, neo-liberal wisdom may demand that a state cuts its bureaucracy at the conclusion of a civil war in order to keep down inflation and achieve international competitiveness. Yet, the political loyalty of particular sections of the population may have been dependent on patronage from the state in terms of employment or access to public goods. Neo-liberal interventions may thus radically alter political bonds and have far-reaching consequences for political participation and stability, and public perceptions of political processes and institutions.

Unsurprisingly, neo-liberalism has attracted immense criticism, particularly in relation to its inability to address poverty and social exclusion. The near deification of the market and corporate power has, according to its critics, reconfigured the balance of power in many states, with public interests being demoted. Harvey (2005: 19) regards neo-liberalism as a 'system of justification and legitimation' aimed at reinforcing 'the capitalist social order'. In this view, it is a potentially authoritarian political model: 'The neoliberal state is necessarily hostile to all forms of social solidarity that put restraints on capital accumulation' (Harvey 2005: 75). While Dorn (1998: 14) regarded 'Chile's free market revolution' of 1973 as 'an example for the rest of Latin America', Harvey (2005: 8–9) was excoriating of Pinochet's US-backed coup on 'little September 11th' and the widespread repression that followed in the name of libertarian ideas. It is true that many champions of neo-liberalism are agnostic about the social and political costs of an unfettered market. Deepak Lal (1998: 68, 70) observed that 'The characteristics of good government are more important than its particular form' and 'it is by no means self-evident ... that Western democracy necessarily promotes a market-friendly culture.' Critics remain unconvinced of the redistributive potential of the market and believe that politics do matter, particularly in relation to political commitments to social inclusion (Harriss 2005b: 228).

Certainly there is a widespread understanding of the potentially pernicious effects of market-led programming in development and peacebuilding contexts (Chua 2004). The problem for critics of neo-liberalism is that their opponents have created a self-reinforcing *system* based on widely accepted norms of efficiency, cost-effectiveness and enterprise. The neo-liberal system is promulgated by corporate interests and international financial institutions, chimes with populist causes, has reconfigured the ethos of public sector institutions (from hospitals to universities) in the developed and developing world, and is in alignment with the strategic interests of leading states. Critics are perfectly correct in pointing to the spectre of 'predatory disaster capitalism' that profits from the misery of others (Klein 2007), but the structures of the contemporary international political economy have been captured by neo-liberal forces and the system seems able to sustain itself (or at least defer or pass on the costs) for the foreseeable future.

The international responses to the global financial crisis that began in 2007 are instructive. Governments all over the world were in thrall to markets and their promise of continuous economic growth (Blinder 2013). But light touch and negligent regulation of markets had given rise to a cannibalistic form of capitalism whereby capital flitted from country to country without responsibility to governments, citizens or any principle other than shareholder value (O'Toole 2010). When the fragility of this house of sand was revealed, governments swung into action. But rather than punish bankers, much of the burden was passed to citizen taxpayers. Banks were bailed out and their debts were nationalized and made the responsibility of the taxpayer (Helleiner 2014; Wolf 2014). The social consequences have been shocking and have helped push the issue of inequality up the political agenda in a number of countries. In 2014, youth unemployment in Spain was 54.3 per cent and in Greece 59 per cent (House of Lords 2014: 16). The economic system, even though it was dysfunctional, was saved. Some additional regulations were introduced but, in general, a neo-liberal system was entrenched.

The predominance of neo-liberalism does not mean that development assistance aimed at emancipating populations, increasing opportunities and promoting redistribution in the developing world has come to an end. Certainly, neo-liberal structures and principles guide much development activity, but the development sector is flourishing. In 2014, worldwide overseas development assistance from the Development Assistance Countries (DAC) was US$135 billion (though in true free market style, much of that was creamed off by Western consultants) (OECD 2015). The United States alone spent almost $33 billion on development aid in 2014, although that amounted to just 0.19 per cent of Gross National Income (GNI). Sweden donates over 1 per cent of GNI. The face of development aid is changing fast though. Regional organizations, oil-rich Gulf and Arab states, and China (through infrastructure development) are now major development and humanitarian actors (IRIN 2014; Campbell and Hofmann 2015: 191–203; Mawdsley 2015: 204–14) are fast becoming major development donors. Increasing emphasis is placed on integrated development interventions, so that the economic, environmental, political and social dimensions of development are interlinked and mutually supporting (Baker 2006). The focus of much development and post-war reconstruction has been on poverty and 'pro-poor growth'. Dealing with poverty is politically unproblematic if economies are growing: the rich and the poor both do well and the rich maintain their comparative advantage. More politically problematic, however, is dealing with inequality. For societies to become more equal, then those with power, money and advantage have to give some of that up. Rich states and international organizations have kept inequality off the agenda by talking about poverty. Yet as the divide between rich and poor becomes ever more apparent, the issue of inequality is harder to suppress (Piketty 2014; Stiglitz 2015).

Sustainability, local participation and ownership, and pro-poor initiatives are a common vein through contemporary development thinking and practice. The UN Millennium Development Goals (MDGs), as agreed by 189 governments in 2000, have crystallized development priorities:

- Eradicate extreme poverty and hunger
- Achieve universal primary education
- Promote gender equality and empower women
- Reduce infant mortality
- Improve maternal health
- Combat HIV/AIDS, malaria and other diseases
- Ensure environmental sustainability
- Develop a global partnership for development.

These were extremely influential as they shaped development policies by international organizations, donor states and international non-governmental organizations (INGOs). A target date of 2015 was set and many international organizations and donor states explicitly mentioned the MDGs in their programme goals. The MGDs were also significant in what they left out: tackling poverty was mentioned but inequality was not, and agriculture was not mentioned. The list was also apolitical and seemed incurious about the political economy that lies behind many developmental and humanitarian problems. There have been some successes, such as cutting global poverty, although this is due, in large part, to China's rapid economic growth. The Millennium Development Goals are due to be replaced by a more ambitious list of Sustainable Development Goals (SDGs). In draft form they are:

1) End poverty in all forms everywhere
2) End hunger, achieve food security and improved nutrition, and promote sustainable agriculture

3) Ensure healthy lives and promote wellbeing for all at all ages
4) Ensure inclusive and equitable quality education and promote lifelong learning opportunities for all
5) Achieve gender equality and empower all women and girls
6) Ensure availability and sustainable management of water and sanitation for all
7) Ensure access to affordable, reliable, sustainable and modern energy for all
8) Promote sustained, inclusive and sustainable economic growth, full and productive employment, and decent work for all
9) Build resilient infrastructure, promote inclusive and sustainable industrialization, and foster innovation
10) Reduce inequality within and among countries
11) Make cities and human settlements inclusive, safe, resilient and sustainable
12) Ensure sustainable consumption and production patterns
13) Take urgent action to combat climate change and its impacts
14) Conserve and sustainably use the oceans, seas and marine resources for sustainable development
15) Protect, restore and promote sustainable use of terrestrial ecosystems, sustainably manage forests, combat desertification and halt and reverse land degradation, and halt biodiversity loss
16) Promote peaceful and inclusive societies for sustainable development, provide access to justice for all and build effective, accountable and inclusive institutions at all levels
17) Strengthen the means of implementation and revitalize the global partnership for sustainable development.

The list is augmented by 169 targets and is interesting because many of the issues it tackles have a political dimension.

Although development activities have been subject to immense faddism since the late 1940s, and although there have been broad shifts in emphasis (from planning to the market-led initiatives), a number of constants are worth noting. The first is that development is still largely a North to South enterprise: intellectually, practically and financially. The second constant is that macro-economic structures are still biased towards the global North. There has been a general trend of wealth and capacity moving eastwards as China, Indonesia, Vietnam and others experience rapid growth. But global economic structures award advantages to already rich states. The third constant factor is that many of the states that were in the most need of development in the 1940s and 1950s (during the peak of development planning optimism) are still grossly undeveloped.

There is no great mystery about the causes and cures of underdevelopment. As Jeffrey Sachs (2008: 6) observed, 'Reaching the MDGs won't take miracles – we know how to keep children alive in malarial regions, we know how to increase food production.' The key issues are linked to political economies that perpetuate inequality. Dealing with inequality (as opposed with simply reducing poverty) requires those that have wealth and capacity to lose some of that. That is not a message that any political leader would find easy to sell.

Although this section has discussed development theory and policy, it is worth noting that populations in the global South have not been sitting back waiting to be developed. Despite a global trading system stacked against them, and often unsympathetic governments, people in Cambodia, Georgia, Uganda and many other countries have simply got on with the tasks of making a living, setting up businesses, consuming, sending their kids to school, migrating and investing in new technology. Governments may have helped with infrastructure and other initiatives, but by and large, people have just got on with the business of life and, in the process, have fuelled economic development (Bøås 2014).

For Indian households, the bicycle became a motorbike and now is becoming a car. For Ugandans the mobile phone network allowed for the electronic transfer of money and so businesses could expand. In Somalia, traders operating through Dubai made contact with counterparts in China and opened up new supply routes. The key point is that aside from macro-economic structures and conditions, people all over the world have used ingenuity, exploited opportunities and have sought to prosper. It is not clear that academic theories of development have recognized this tendency of people to get on with things, to have their own economic survival mechanisms, and to regard formal development projects and programmes as often irrelevant to their circumstances.

The evolution of conflict theory

It is worth noting that for many writers, violent conflict was regarded as a given (a mere by-product of Great Power interplay or the consequence of natural inclinations of humankind), and its causes were not subject to serious scrutiny. Military history, which depicted war as 'a deplorable necessity' (Creasy 1876: xi), thrived, but did little to advance our understanding of the precipitants of conflict, or its wider social or cultural impacts. More considered deliberations on the nature and causes of warfare occupied scholars from Sun Tzu (544–496BC) and Thucydides (460–395BC) to Saint Augustine (354–430) and Hugo Grotius (1583–1645), but their work did not constitute a united or recognizable field of conflict studies (Jacoby 2008: 8–12). Real world 'traumas' such as the First World War (over 35 million casualties, the industrialization of warfare and the militarization of societies), or the unleashing of nuclear weapons gave renewed impetus to those attempting to systematize knowledge of conflict. For example, Lewis Fry Richardson (1950), a Quaker conscientious objector who served with the Friends' Ambulance Unit during the First World War, sought to apply mathematical analysis to warfare, particularly the propensity of armament programmes to lead to conflict. Yet Richardson, and other pioneers of the study of conflict as a generic phenomenon such as Quincy Wright (1942) and Georg Simmel (1955), constituted a distinct minority. Indeed, many of those who pursued 'peace studies' were derided as cranks, cowards, unpatriotic or 'religious nuts' (Rooney 2000: 16).

It has not been until relatively recently that attempts to find general theories of conflict reached the academic mainstream (Azar 1990: 5). The subdiscipline of international relations, for example, has traditionally been concerned with wars between states and showed little inclination to investigate sub-state conflicts. It was not until the 1950s and 1960s that a recognizably modern strain of conflict research emerged. Miall (2007: 27) notes how much of this research attempted 'to capture the generic characteristics of conflict', stripping conflicts of their context in order to better examine the relationships between actors and the dynamics of conflict processes. Much of this research was influenced by game theory, behaviouralism and the application of social psychology to conflict (Axelrod 1990; Boulding 1990: 37–8). Rational choice and bargaining theory approaches to the study of conflict are now particularly popular (especially among North American and Scandinavian academics) and are prominent in leading academic journals of peace and conflict. These approaches have helped bring evidence and data into arguments that were often based on assertion (Regan 2013). Comprehensive databases of conflicts such as the Conflict Data Program at Uppsala, the Peace Accords Matrix or the Armed Conflict Database have allowed scholars and policymakers to study issues such as the optimal time to intervene in a conflict, or the chances of particular types of conflicts having negotiated or violent outcomes.

Over time, more holistic understandings of conflict have emerged among policymakers and academics. This has been complemented by the 'mainstreaming' of peace and conflict research into a number of academic disciplines, with the establishment of research institutes, specialist journals and a greater acceptance of the systematic study of conflict in policymaking circles. There has been something of an intellectual emancipation of peace and conflict. No longer is it the preserve of a few pacifists and the closed shop of military and foreign policy officials. Indeed, after decades of being on the margins, the study of peace and conflict has even become popular. Christopher Mitchell (1994: 128) recalls how, as a reaction to the rash of civil wars in the 1990s, 'a range of scholars … discovered that they have "really" been doing conflict resolution "all along".' 'Ex-strategic theorists, military security experts, Sovietologists and area specialists' suddenly turned their attentions to the problems of civil war. As will be outlined in Chapter 1, the rush to 'explain civil war' has variously focused on the importance of identity, regional factors and econometric indicators. The modern evolution of conflict studies has reflected real world conditions. The post-Cold War upsurge in conflicts, greater interventionism by international organizations and international NGOs (INGOs), and the impact of globalization on public awareness of conflicts on the other side of the planet, demanded a better understanding of conflict. The failure (or limited success) of some immediate post-Cold War international interventions demanded a further refinement of our understanding of conflict.

Specialist subfields in the study of conflict developed in the 1990s and beyond, with the comparative study of negotiated peacemaking processes (Darby and Mac Ginty 2000), gender and security sector reform (Özerdem 2010) gaining particular attention. A number of conflict resolution 'gurus', or respected practitioners such as John Paul Lederach, Roger Fisher, William Ury or Ben Hoffman, also gained greater prominence (Fisher and Ury 1991; Lederach 1995; Hoffman 2007). There was a growing consensus on the multidimensional nature of conflict, its increasingly transnational nature in a globalized context, the need to see beyond conflict manifestations to examine conflict causes, and – crucially – the importance of development issues in explaining conflict (Azar 1990: 2). Research on conflict has also recognized the protracted nature of many conflicts, the tendency of violent conflicts to reignite following periods of calm, and the difficult, long-term and costly nature of post-peace-accord peacebuilding. Indeed, 'no war, no peace' situations have become common in which parties agree to a ceasefire but fail to push for a comprehensive peace process (for example, in Sri Lanka, Israel–Palestine and Colombia at various times in the 1990s and 2000s), or reach a peace agreement but fail to move towards a widespread reconciliation (for example, Northern Ireland or Lebanon) (Mac Ginty 2006).

The increased research on peace and conflict was often part of desperate attempts to understand conflict. Much research has been sponsored by governments, international organizations and development agencies. Theories of peace and conflict have largely bifurcated into problem-solving and critical approaches (Cox 1981; Pugh 2013). The problem-solving approach is interested in remedying conflict manifestations. It is often policy orientated and is attractive in that 'something is being done'. The critical approach, on the other hand, is less easily satisfied. It concentrates on the structural factors that undergird violent conflict and is suspicious of peace-support interventions that may staunch fighting but leave power and economic relations largely unchanged (Schmid 1968). The critics accuse the problem-solvers of being incurious about the factors that start and maintain conflict. The problem-solvers accuse the critics of constantly complaining but never getting around to doing anything. In this view, the system of intervention is far from perfect but we must work with what we have (Paris 2010).

Importantly, there has been an elision of many international strategies to deal with conflict and underdevelopment. Major documents by international organizations show an awareness of how conflict and development often have a symbiotic relationship and should be tackled together. Many of the same strategies that are deployed in societies emerging from civil war can also be found in societies free from civil war but suffering from underdevelopment. Development strategies in societies emerging from civil war might be modified so as to be 'conflict sensitive', but they essentially amount to the same thing. Over the past few decades there has been increased emphasis on the 'local ownership' of 'solutions' on the understanding that if local communities feel invested in a programme or project then it is likely to be more effective and sustainable. Arguments rage about the extent to which this 'local turn' in development and peacebuilding planning is rhetoric or denotes a more fundamental shift (Richmond 2012): the promotion of the orthodoxy of neo-liberalism, 'good' governance reforms, and the use of aid conditionality or selectivity to encourage conformity.

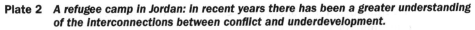

Plate 2 *A refugee camp in Jordan: in recent years there has been a greater understanding of the interconnections between conflict and underdevelopment.*

Alongside the critical strain to the literature on peace, conflict and development, other scholars and policymakers have been vexed at how to 'win' the War on Terror and subjugate opponents in Afghanistan, Iraq and elsewhere. The United States and UK increasingly see development and post-war reconstruction as a key means of winning 'hearts and minds' in the support of their conquest of Iraq and Afghanistan. Thus we have seen the securitization of development or the incorporation of development into security strategies. This has been most visible in the provincial reconstruction teams at work in Afghanistan and Iraq whereby soldiers become 'armed humanitarians', with obvious consequences for ideas of neutrality and impartiality which some associated with humanitarianism. Critics say that Western states are more interested in order and security than development and reconstruction. For example, in early 2007, the United States committed an additional 28,829 troops to its war effort in Iraq. Over 21,000 of these were combat troops and just 129 were tasked with provincial reconstruction (*Guardian* 2007). Such militarized development results in highly contradictory international interventions: in less than a year British troops fired over four million bullets in Afghanistan *and* spent over £100 million in development assistance (Harding 2008).

A conflict-specific nomenclature has developed, with a battery of prefixes and suffixes fine-tuning our understanding of the concepts of peace and conflict. Thus 'peacekeeping' denotes traditional United Nations (UN) 'blue helmet' troop deployments, usually to police an agreed separation between antagonists. 'Peacemaking' suggests a more active form of intervention or activity, possibly including force to compel parties to negotiate or make concessions. 'Peacebuilding' was originally regarded as an activity that occurred after a peace accord was reached in order to support that accord. It covered a range of political, social, economic and cultural activities and so was cognisant of the links between conflict, peace and development. The strict delineation of peacebuilding as an activity that occurs after a violent conflict ceased no longer holds. 'Peace implementation' usually refers to both the fulfilment of the provisions of a peace accord and attempts to provide an environment (for example, security, minority return, a tax base to fund social spending) to enable the implementation of the terms of the accord. 'Peace fixing', if we may be permitted to coin a new term, is the common activity of internal and external parties to a peace accord returning to the negotiating table to modify the accord to react to problems. In the post-9/11 period, there has been a slight draw back from the use of the word 'peace' by a number of leading states as they promote security agendas (often under the term 'stabilization').

There has also been a transition in the suffixes used in relation to conflict. 'Conflict resolution' was overtaken by 'conflict management' when some critics suggested that conflicts could not be definitively resolved. Instead of finding solutions, societies had to recognize the integral nature of conflict to human societies and thus find non-violent ways of managing the conflict. But the term 'conflict management' was criticized for assuming that some actors would be managers (often powerful or well-resourced external actors), while other actors would be managed (often indigenous, less powerful actors). The term 'conflict transformation' is now current (doubtless to be supplanted by another term in the coming years). Conflict transformation again recognizes that conflict is part and parcel of human existence, but aims at transforming relationships between individuals, groups and institutions from destructive to constructive bonds.

Peacebuilding

Peacebuilding, as a catch-all term that is much used in policy documents, and indeed this book, deserves elaboration. The term is often
more 'described than defined', comes with or without hyphens and it
still does not appear in the *Oxford English Dictionary* (Holt 2011: 32;
see also Jeong 2005). Johan Galtung describes his version of 'peacebuilding' as 'building structural and cultural peace') possibly as vague
as his core idea of 'structural violence'; John Paul Lederach sees it
encompassing the 'full array of processes, approaches and stages
needed to transform conflict toward more sustainable, peaceful relationships' (Galtung 1996; Lederach 1997, both quoted in Holt 2011: 32–3).

Consequently we might say that it has changed in line with historical
context, both in the 'target' areas where conflict is a problem, but also
in the institutional reaction to these areas, for the UN itself is also
subject to a certain amount of definitional imprecision when trying to
explain what peacebuilding aims to do. But what is common to all
peacebuilding approaches is a realization that violent conflict (not
conflict *per se*) 'is inflicting immense damage on the societies and
economies of the developing world'. Collier and Hoeffler have been
quoted as estimating that a 'typical' developing country civil war
inflicts at least US$64.2 billion of economic and social damage
(Collier and Hoeffler 2004, quoted in Addison 2005: 406).

Originally, many regarded peacebuilding as a post-conflict activity –
after violence had ceased. The chronic nature of conflicts (for example,
in parts of the Horn of Africa) means that it is difficult to maintain an
argument that actors should wait until conflict ends before trying any
ameliorative effort. Thus peacebuilding has become a generic term for
conflict response interventions before, during and after violent conflict.
An interest in conflict prevention in the 1990s (Boutros-Ghali 1992;
UN 2000; Carnegie 1997) has been superseded by international
concern for resilient states and communities, able to withstand
destablizing pressures. There is difficulty identifying where peacebuilding, conceptually and practically, begins and ends. Certainly there
is consensus that conflict demands multidimensional responses, and
that local voices have to be incorporated into the design and implementation of the response (Addison 2005; Richmond 2006: 74).

The ultimate aim of peacebuilding can be regarded as 'sustainable' or
even a 'stable' peace; classically defined by Karl Deutsch as being

when there is the 'real assurance that the members of [a] community will not fight each other physically, but will settle their disputes in some other way' (Deutsch quoted in Kacowicz et al. 2000: xi). This has been refined as where '[political] leaders craft political institutions that will sustain the peace and foster democracy in ethnically divided societies after conflicts such as civil wars' (Roeder and Rothchild 2005: 3). These ambitions beg a lot of questions, particularly regarding where power might lie, and the 'rights' of Western states to intervene in societies and states thousands of miles away (Chabal and Daloz 1999, and 2006: vii).

Fundamentally, and this is a major theme throughout this book, peace-building raises questions about the balance between local, national, regional and international actors. In many circumstances, national, regional and international actors may have power and capacity, but they may lack legitimacy among local populations who have to endure the privations of conflict, de-development and under-development. Obviously there are immense practical issues involved here, but crucial to the issue is how we see the world. Do we accept terms like 'peacebuilding', 'stability' and 'international community' uncritically, or do we take them apart and see what they are made of? This book hopes to contribute to the process of taking apart ideas, and practices. Patrick Chabal's challenge still stands. He said that that Western development economists and political commentators have always seen 'Africa [as the] product of its own imagination rather than that of a serious interest in what actually happens on the continent' (Chabal in Werbner and Ranger, 1996: 36) The challenge is therefore how Western-dominated intergovernmental organizations (IGOs) and states (our 'hardware') can change their 'software' to accommodate the different realities of developing countries with vastly different cultures and problems than our own.

Structure of the book

Chapter 1 advances our understanding of conflict and development by examining the links between poverty, profit and violent conflict. It outlines the main theories that explain the escalation and maintenance of conflict and discusses the political economy of violence whereby conflict becomes a profitable activity and therefore – in some cases – sustainable. The chapter also examines the connection between the presence and exploitation of natural resources and conflict. Chapter 2 provides an overview of the institutional architecture that shapes contemporary peace and development: states, the market, international organizations and international financial institutions. Contemporary conflict and development takes place in a context of complex multilateralism and hyper-globalization in which multiple actors can be involved in the same conflict and/or development process, often using proxies and sometimes acting in contradictory ways. The chapter seeks to make sense of the institutional architecture of conflict and development using the debates on the 'liberal peace'.

In contrast to Chapter 2 and its focus on top–down institutions, Chapter 3 deals with people, as citizens, the displaced, victims, antagonists, consumers of public goods, men or women, unified in civil society or fragmented into particularistic groups. The elixir for those who wish to promote conflict, peacebuilding or development is popular participation as a means of legitimizing their enterprise. This chapter will explore the impact of conflict and development on people, and the strategies that ethnic entrepreneurs, political leaders, aid agencies and others use to mobilize groups and legitimize their activities.

Chapters 4, 5 and 6 are essentially intended to show how policymakers and academic students of conflict alike have tried to conceptualize the problems that arise after conflicts have arisen and, in some cases, 'ended'. We have summed it up as 'transitions' from conflict and war to a kind of 'peace' (Organization for Economic Cooperation and Development (OECD) 2007: Foreword). Chapter 4 looks at thinking on conflict resolution (though we prefer the term 'transformation') and the techniques and methods that have emerged since the end of the Cold War to try to damp down conflicts and deal with their psychological, social and other 'deficits', as Miall et al. (1999) and Ramsbotham et al. (2005) have put it. This will encompass an examination of different approaches to ending conflict, including newer approaches such as truth and reconciliation commissions. Chapter 5 looks at the evolving notion of 'reconstruction' after wars, and in particular takes issue with the belief that we can have a 'one-size-fits-all' approach to such efforts. It will in particular flesh out what we called earlier the 'increasingly technocratic [nature of] development and peace-support interventions'. The alphabet soup of reconstruction now includes acronyms like DDR (disarmament, demobilization and reintegration) and SSR (security sector reform) that have to be understood by any neophyte (or advanced) student of conflict and development. We will attempt to sieve the soup in such a way as to make clear both the underlying rationale for such terms and practices, but also their implementation. In this chapter we will also look in more detail at some of the human and structural problems involved in 'reconstructing' a society. What are the problems to do with the health of the population that need to be considered after a war, for example? Finally, Chapter 6 comprises an overview of the difficulties of using 'aid' as a panacea during, but mainly after, war. The focus here will be on those delivering aid, and the reaction of those to whom it is delivered, which is not always positive. We consider that this dilemma is one of the keys to understanding what is both wrong, and right, with current development policies in the developing world. Along the way, text boxes are used to illustrate points. Chapters will end with summaries, discussion questions and suggested further reading. Since many of the organizations and issues mentioned in the chapters have content-rich websites, e-resources are annotated at the end of each chapter.

Summary

- Theories of conflict and development have largely evolved in isolation from one another.
- More recently, there has been a greater cross-fertilization between ideas and policy approaches to conflict and underdevelopment.
- Particular worldviews have dominated thinking about conflict and development. At the moment, neo-liberal economic ideas influence development thinking and, in the aftermath of 9/11, security and order are playing a prominent role in peace promotion.
- More complex understandings of conflict and development have taken root in the policy and scholarly worlds.
- There is a growing realization that there are no 'quick fixes'.

Discussion questions

- How could 'success' in development be defined?
- How can we make sure that development and peace-support interventions do not become new forms of imperialism?
- Consider if there are realistic alternatives to neo-liberal development models.

Further reading

Excellent surveys of the evolution of development theory and practice can be found in C. Leys (1996) *The Rise and Fall of Development Theory*, Oxford: James Currey; and U. Kothari (ed.) (2005) *A Radical History of Development Studies: Individuals, institutions and ideologies*, London: Zed. A neo-liberal assault on Keynesian or state-led development strategies can be found in J. Dorn, S. Hanke and A. Walters (eds) (1998) *The Revolution in Development Economics*, Washington, DC: Cato Institute. O. Ramsbotham, T. Woodhouse and H. Miall (2011) *Contemporary Conflict Resolution: The prevention, management and transformation of deadly conflicts*, 3rd edition, Cambridge: Polity, provide an excellent survey of literature and approaches to conflict. Major edited volumes with contributions from leading authors on peacebuilding and humanitarianism are: R. Mac Ginty (ed.) (2013) *Routledge Handbook of Peacebuilding*, London: Routledge; and R. Mac Ginty and J. Peterson (eds) (2015) *Routledge Companion to Humanitarian Action*, London: Routledge.

Useful websites

A goldmine of material on conflict and conflict transformation can be found at Beyond Intractability (www.beyondintractability.org/), Open Democracy (https://www.opendemocracy.net/), International Alert (http://www.international-alert.org/) and International Crisis Group (http://www.crisisgroup.org/). The Humanitarian Practice Network provides online reports on development and humanitarian issues: www.odihpn.org/. See also the Global Humanitarian Assistance website, particularly its annual reports: http://www.globalhumanitarianassistance.org/.

UN documents on the Sustainable Development Goals can be found here: https://sustainabledevelopment.un.org/topics/sustainabledevelopmentgoals. The *Guardian* newspaper has an excellent resource of articles and data on development: http://www.theguardian.com/global-development.

References

Addison, T. (2005) Conflict and Peace Building: Interactions between Politics and Economics. In Addison, T. (ed.) Special Issue of *Round Table: The Commonwealth Journal of International Affairs* 94(381): 405–11.

Amnesty International (2013) China, Annual Report 2013: The state of the world's
  human rights. Available at: http://www.amnesty.org/en/region/china/report-2013
  (accessed 1 February 2015).
Axelrod, R. (1990) *The Evolution of Cooperation*. London: Penguin.
Azar, E. (1990) *The Management of Protracted Social Conflict: Theory and cases*.
  Aldershot: Dartmouth.
Baker, S. (2006) *Sustainable Development*. London: Routledge.
Bauer, P. (1998) The Disregard of Reality. In Dorn, J., Hanke, S. and Walters, A.
  (eds) *The Revolution in Development Economics*. Washington, DC: Cato Institute,
  pp. 25–39.
Benn, T. (2001) Hansard: record of parliamentary debates, column 510, 22 March.
  Available at: http://www.publications.parliament.uk/pa/cm200001/cmhansrd/
  vo010322/debtext/10322-13.htm (accessed on 1 February 2015).
Blinder, A.S. (2013) *After the Music Stopped: The financial crisis, the response and
  the work ahead*. London: Penguin.
Bøås, M. (2014) *The Politics of Conflict Economies: Miners, merchants and warriors
  in the African borderland*. London: Routledge.
Boulding, K. (1990) Future Directions in Conflict and Peace Research. In Burton, J.
  and Dukes, F. (eds) *Conflict: Readings in management and resolution*. London:
  Macmillan, pp. 35–47.
Boutros-Ghali, B. (1992) *An Agenda for Peace Preventive diplomacy,
  peacemaking and peace-keeping*. Report of the Secretary-General pursuant to
  the statement adopted by the Summit Meeting of the Security Council on 31
  January 1992.
Campbell, S. and Hofmann, S. (2015) Regional Humanitarian Organisations. In Mac
  Ginty, R. and Peterson, J. (eds) *Routledge Companion to Humanitarian Action*.
  London: Routledge, pp. 191–203.
Carnegie Commission on Preventing Deadly Conflict (1997) *Preventing Deadly
  Conflict*. New York: Carnegie Corporation of New York.
Chabal, P. and Daloz, J.-P. (1999) *Africa Works: Disorder as a political instrument*.
  Oxford: James Currey.
Chabal, P. and Daloz, J.-P. (2006) *Culture Troubles: Politics and the interpretation
  of meaning*. London: Hurst.
Chua, A. (2004) *World on Fire: How exporting free market democracy breeds ethnic
  hatred and global instability*. New York: Anchor.
Clapham, C. (1987) Revolutionary socialist development in Ethiopia. *African Affairs*
  86(343): 151–65.
Cohen, S.B. (2014) The challenging dynamics of global north–south peacebuilding
  partnerships: practical stories from the field. *Journal of Peacebuilding and
  Development* 9(3): 65–81.
Collier, P. (2000b) *Economic Causes of Civil Conflict and their Implications for
  Policy*. Washington, DC: World Bank.
Collier, P. and Hoeffler, A. (2004). Greed and grievance in civil war, *Oxford
  Economic Papers* 56(4): 563–95.
Cox, R.W. (1981) Social forces, states and world orders: beyond international
  relations theory. *Millennium* 10: 126–56.
Creasy, E. (1876) *The Fifteen Decisive Battles of the World: From Marathon to
  Waterloo*. London: Bentley.

Darby, J. and Mac Ginty, R. (eds) (2000) *The Management of Peace Processes*. Basingstoke: Macmillan.

Davenport-Hines, R. (2015) *Universal Man: The seven lives of John Maynard Keynes*. London: Collins.

Dorn, J. (1998) Competing Visions of Development Policy. In Dorn, J., Hanke, S. and Walters, A. (eds) *The Revolution in Development Economics*. Washington, DC: Cato Institute, pp. 1–21.

Endres, A. (2012) *Architects of the International Financial System*. London: Routledge.

Fisher, R. and Ury, W. (1991) *Getting to Yes: Negotiating agreement without giving in*. London: Penguin.

Frank, A.G. (1967) *Capitalism and Underdevelopment in Latin America*. New York: Monthly Review Press.

Galbraith, J.K. (1964) *Economic Development*. Cambridge, MA: Harvard University Press.

Galtung, J. (1996) *Peace by Peaceful Means: Peace and conflict, development and civilization*. London: Sage.

Guardian (2007) Iran's secret plan to force the US out. *Guardian*, 22 May.

Gurr, T. (1970) *Why Men Rebel*. Princeton, NJ: Princeton University Press.

Harb, C. and Saab, R. (2014) *Social Cohesion and Intergroup Relations: Syrian refugees and Lebanese nationals in the Bekka and Akkar*. London: Save the Children.

Harding, T. (2008) A year in Helmand: 4m bullets fired by British. *Daily Telegraph*, 13 January.

Harriss, J. (2005a) Great Promise, Hubris and Recovery: A Participant's History of Development Studies. In U. Kothari (ed.) *A Radical History of Development Studies: Individuals, institutions and ideologies*. London: Zed, pp. 17–46.

Harriss, J. (2005b) Do Political Regimes Matter? Poverty Reduction and Regime Differences across India. In Houtzager, P. and Moore, M. (eds) *Changing Paths: International development and the new politics of inclusion*. Ann Arbor, MI: University of Michigan Press, pp. 204–32.

Harvey, D. (2005) *A Brief History of Neoliberalism*. Oxford: Oxford University Press.

Helleiner, E. (2014) *The Status Quo Crisis: Global financial governance after the 2008 meltdown*. Oxford: Oxford University Press.

Hoffman, B. (2007) *The Peace Guerilla Handbook*. York: Newmathforhumaity.

Holt, S. (2011) *Aid, Peacebuilding and the Resurgence of War: Buying time in Sri Lanka*. London: Palgrave Macmillan.

Homans, C. (2012) Energy Independence: A short history. *Foreign Policy*, 3 January. Available at: http://foreignpolicy.com/2012/01/03/energy-independence-a-short-history/ (accessed on 1 February 2015).

House of Lords (2014) *Youth Unemployment in the EU: A scarred generation?* London: The Stationery Office.

International Development Committee (2006) *Conflict and Development: Peacebuilding and post-conflict reconstruction*. Sixth Report of the Session 2005–06, Vol. 1. London: The Stationery Office.

IRIN (2014) Turkey's ambition as a rising donor. IRIN, 27 March. Available at: http://www.irinnews.org/report/99848/turkey-s-ambitions-as-a-rising-donor (last accessed on 16 September 2015).

Jacoby, T. (2008) *Understanding Conflict and Violence: Theoretical and interdisciplinary approaches*. London: Routledge.

Jeong, H.-W. (2005) *Peacebuilding in Post-Conflict Societies: Strategy and process*. Boulder, CO and London: Lynne Rienner.

Kacowicz, A.M., Bar-Siman-Tov, Y., Elgström, O., Jerneck, M. (eds) (2000) *Stable Peace Among Nations*. Lanham, MD: Rowman and Littlefield.

Kaldor, M. (2007) *New and Old Wars: Organized violence in a globalized world*, 2nd edn. London: Polity.

Keynes, J.M. (1920) *The Economic Consequences of Peace*. London: Macmillan.

Klein, N. (2007) *The Shock Doctrine: The rise of disaster capitalism*. London: Penguin.

Kothari, U. (2005) A Radical History of Development Studies: Individuals, Institutions and ideologies. In Kothari, U. (ed.) *A Radical History of Development Studies: Individuals, institutions and ideologies*. London: Zed, pp. 1–13.

Lal, D. (1998) The Transformation of Developing Economies: From Plan to Market. In Dorn, J., Hanke, S. and Walters, A. (eds) *The Revolution in Development Economics*. Washington, DC: Cato Institute, pp. 55–74.

Lederach, J.P. (1995) *Preparing for Peace: Conflict transformation across cultures*. Syracuse, NY: Syracuse University Press.

Lederach, J.P. (1997) *Building Peace: Sustainable reconciliation in divided societies*. Washington, DC: United States Institute of Peace Press.

Leys, C. (1996) *The Rise and Fall of Development Theory*. London: James Currey.

Mac Ginty, R. (2006) *No War, No Peace: The rejuvenation of stalled peace processes and peace accords*. London: Palgrave.

Mawdsley, E. (2015) 'Non-DAC' Humanitarian Actors. In Mac Ginty, R. and Peterson, J. (eds) *Routledge Companion to Humanitarian Action*. London: Routledge, pp. 204–14.

Meier G.M. and Rauch, J.E. (2000) *Leading Issues in Economic Development*. New York: Oxford University Press.

Miall, H. (2007) *Emergent Conflict and Peaceful Change*. Basingstoke: Palgrave Macmillan.

Miall, H., Ramsbotham, O., Woodhouse, T. (1999) *Contemporary Conflict Resolution*, 1st edn. London: Polity.

Mitchell, C. (1994) Conflict Management. In Groom, A.J.R. and Light, M. (eds) *Contemporary International Relations: A guide to theory*. London: Pinter, pp. 128–41.

OECD (2007) *Handbook on 'Security Sector Reform'*. Paris: OECD.

OECD (2015) Development aid stable in 2014 but flows to poorest countries still falling. 8 April. Available at: http://www.oecd.org/dac/stats/development-aid-stable-in-2014-but-flows-to-poorest-countries-still-falling.htm (accessed on 8 September 2015).

O'Toole, F. (2010) *Ship of Fools: How stupidity and corruption sank the Celtic Tiger*. New York: PublicAffairs.

Özerdem, A. (2010) Insurgency, militias and DDR as part of security sector reconstruction in Iraq: how not to do it, *Disasters*. 34(1): S40–S59.

Paris, R. (2010) Saving liberal peacebuilding. *Review of International Studies* 36(2): 337–65.

Pettersson, T. and Wallensteen, P. (2015) Armed conflicts, 1946–2014. *Journal of Peace Research* 52(4): 536–50.

Piketty, T. (2014) *Capital in the Twenty-First Century*. Harvard, MA: Harvard University Press.

Pugh, M. (2013) The Problem-solving and Critical Paradigms. In Mac Ginty, R. (ed.) *Routledge Handbook of Peacebuilding*. London: Routledge, pp. 11–24.

Ramsbotham, O., Woodhouse, T., Miall, H. (2005) *Contemporary Conflict Resolution*, 2nd edn. London: Polity.

Regan, P.M. (2013) Quantitative Approaches. In Mac Ginty, R. (ed.) *Routledge Handbook of Peacebuilding*. London: Routledge, pp. 183–93.

Richardson, L.F. (1950) *The Statistics of Deadly Quarrels*. Pittsburgh, PA: Boxwood.

Richmond, O. (2006) The problem of peace: understanding the 'liberal peace'. *Conflict, Security and Development* 6(3): 291–314.

Richmond. O. (2012) Beyond local ownership in the architecture of international peacebuilding. *Ethnopolitics* 11(4): 354–75.

Roeder, P.G. and Rothchild, D. (eds) (2005) *Sustainable Peace: Power and democracy after civil war*. Ithaca, NY and London: Cornell University Press.

Rooney, A. (2000) *My War*. New York: PublicAffairs.

Rostow, W.W. (1960) *The stages of economic growth: A non-communist manifesto*. Cambridge: Cambridge University Press.

Sachs, J. (2008) Promises, promises. *Developments* 40: 6.

Schmid, H. (1968) Peace research and politics. *Journal of Peace Research* 5: 217–32.

Scholey, P. (2006) Peacebuilding Research and North–South Research Relationships: Perspectives, Opportunities and Challenges. In MacLean, S., Black, D. and Shaw, T. (eds) *A Decade of Human Security: Global governance and new multilateralisms*. Aldershot: Ashgate, pp. 179–92.

Shaw, T., MacLean, S., Black, D. (2006) Introduction: A Decade of Human Security: What Prospects for Global Governance and New Multilateralisms? In MacLean, S., Black, D. and Shaw, T. (eds) *A Decade of Human Security: Global governance and new multilateralisms*. Aldershot: Ashgate, pp. 3–18.

Simmel, G. (1955) *Conflict: The web of group affiliations*. New York: Free Press.

Stiglitz, J. (2015) *The Great Divide*. New York: Allen Lane.

Thomas, A. (2006) Reflections on Development in a Context of War. In Yanacopulos, H. and Hanlon, J. (eds) *Civil Peace, Civil War*. Milton Keynes: Open University Press, pp. 185–205.

Troy, G. (2007) *Morning in America: How Ronald Reagan reinvented the 1980s*. Princeton, NJ: Princeton University Press.

UNDP (1994) *Human Development Report*. Oxford: Oxford University Press.

United Nations (2000) *Report of the Panel on United Nations Peace Operations*. New York: United Nations, A/55/305.

United Nations Development Programme (2006) *Human Development Report 2006: Beyond Scarcity – Power, poverty and the global water crisis*. New York: UNDP.

Werbner, R. and Ranger, T. (eds) (1996) *Postcolonial Identities in Africa*. London: Zed.

Win, E. (2007) Not Very Poor, Powerless or Pregnant: The African Woman Forgotten by Development. In Cornwall, A., Harrison, E. and Whitehead, A. (eds) *Feminisms in Development: Contradictions, contestations and challenges*. London: Zed, pp. 79–85.

Wolf, M. (2014) *The Shift and the Shocks: What we've learned and what we still have to learn from the financial crisis*. London: Penguin.

World Bank (2013) 'GDP growth (annual %)', World Bank indicators. Available at: http://data.worldbank.org/indicator/NY.GDP.MKTP.KD.ZG (accessed on 1 February 2015).

Wright, Q. (1942) *A Study of War*. Chicago, IL: Chicago University Press.

Poverty, profit and the political economy of violent conflict

Introduction

Among many others, there are two basic views on the economics of violent conflict. In the first view, economic development offers a ladder out of conflict. This ladder operates at the international and individual levels. States that trade together have an interest in stability and continued prosperity and so have no interest in the disruptions that come with war. Similarly, this thinking goes, individuals have an interest in peace. Individuals are likely to behave rationally and so would personally avoid conflict and encourage their political leaders, by voting for moderate parties, to avoid conflict. Rational individuals would see economic development as a route out of conflict. In short, 'free markets made free men' and free men would not be foolish enough to become involved in war (Mandelbaum 2002).

The second view on the economics of conflict points towards the predatory nature of capitalism and how the deep inequalities it causes can fan the flames of conflict. This view of economics as a cause of violent conflict also points towards the profiteering of warriors. Whether arms manufacturers, roadside bandits levying a war 'tax', or those running blood diamond or oil-smuggling operations in war zones, there are hefty profits to be made from war. In such cases, peace is a threat to livelihoods.

So, opinion is polarized on the role of economics (saviour or villain?) in the outbreak of civil war. This chapter will examine the often contradictory literature on conflict and development, using examples to illustrate the linkages between conflict, poverty and profit. The chapter is divided into five sections. The first section examines the various economic and development-related arguments on the causation and escalation of violent conflict. The second section delves more deeply into the complex relationship between economics and civil war. Section three examines the factors behind conflict maintenance and the peculiar economic dynamics that sustain war. An anthropological lens is particularly useful in illustrating the ways in which individuals, communities, armed groups, businesses and states variously prosper, starve or 'get by' during war. The fourth section focuses on corruption and considers how the term is often used selectively and is unhelpful as an analytical category. The fifth section concentrates on the resource environments often found in the sites of violent conflict and considers the extent to which the presence, absence and distribution of resources such as diamonds or water can fuel or calm violent conflict.

Conflict causation and escalation

There is no universal theory of conflict causation. The sheer variety of conflict actors, environments and dynamics makes impossible a general theory of conflict or, indeed, of conflict management. Each conflict has a peculiar 'conflict DNA'. This may be a partial match with other conflicts, but will contain factors specific only to that conflict. Conflicts have multiple causes that interact in highly specific ways according to the context. They can have primary causes that take precedence over secondary causes, but the variegated nature of human polities, economies and societies means that a single factor cannot spark a conflict. Different factors will have different weight at different stages of a conflict trajectory. For example, a single atrocity or grievance may prove inflammatory at the outbreak of a civil war, but it becomes overtaken by other factors that sustain the conflict in the longer term. The task for the conflict analyst is first to identify conflict causes (plural) and then establish the connections between them. This is rarely an easy task, since the public rhetoric used by protagonists may mask truer motives, or because protagonists (such as the Lord's Resistance Army in northern Uganda or state-sponsored militias in Darfur) may be secretive and offer few public clues as to their motivations.

Complicating matters even further is that antagonists operating in the same conflict may have very different motivations. This applies within and between groups. Consider the Israeli–Palestinian conflict. At the group level, recruits may be motivated to join Palestinian militant organizations for any number of reasons: spiritual fulfilment, to avenge a grievance, peer pressure, family tradition, belief in their political and strategic aims, to boost personal esteem or to earn an income. Israelis may join the Israeli Defence Force because of a desire to defend their state and community, to gain new skills and further their career, to continue family tradition and fulfil conscription obligations or to avoid being branded deviant or cowardly by not joining up. So there is a cacophony of motivations swirling around the conflict area. Some of these may be economic but many are not. Trying to construct an intelligible narrative thread out of all of these motivations is a difficult task, but one that is necessary for the conflict analyst. Although Palestinian militants and the Israeli Defence Force share the same 'battlefield' and pledge themselves to the destruction of the other, their conflicts have very different aims. For Israel, the conflict is largely justified in terms of defence and security. For Palestinian militants, it is about righting grievances and promoting a religiously inspired worldview. The conflict analyst is faced with a bewildering array of 'evidence' and, ultimately, must make a judgement call on which factors they believe to be most significant in the escalation and maintenance of a conflict. Box 1.1 illustrates, with reference to Colombia, how multiple factors compete for the attention of the analyst. Rather than a science, conflict analysis is an art and involves human – and therefore potentially frail – judgement.

Given that the focus of this book is on conflict and development, the bulk of this section will dwell on development and economic-related explanations for conflict causation and maintenance. Yet there are entire subfields of literature on conflict causation that make little reference to development, economics, profit or poverty. Non-economic or non-development-related explanations for the outbreak of violent conflict include the often overlapping categories of biological disposition (Simmel 1955), psychology (Tajfel 1978), religion (Appleby 2000), identity (Sen 2006), ethnicity (Connor 1994; Young 2003), nationalism, ideology, history and ancient hatreds, bad neighbours, manipulative leaders (Brown 1997), the security dilemma (Posen 1993), cultural dysfunction (Kaplan 1994), the nature of the state (Tilly 1985) and incompatible worldviews (Huntington 1998). Many of these explanations regard economic factors as contingent, or providing a context in which the primary factor operates. Thus, for example, an ethnic group may develop an elaborate self-narrative of grievance, how its rights are denied, and how it is distinct from other identity groups in the context of declining economic conditions. These conditions may provide the backdrop or even tipping point for

Box 1.1

Colombia: a confusing conflict stratum

A single-word explanation is often given for the long-running war in Colombia: drugs. Drug money fuels both the legal and illegal economies with anti-state guerrillas, pro-state paramilitaries and elements of the state all implicated in the drugs trade. In 2014, Colombia was responsible for 43 per cent of world coca production, an industry that was worth $10 billion annually (Kaplan 2014). But scratch the surface and a more complex conflict stratum is revealed. Certainly the drugs trade is important, but its primary significance has been in maintaining the conflict once it had already begun and in creating a political economy of war. It was not until late 2012 that serious peace negotiations between the government and the main rebel group began in Cuba. At the heart of the conflict has been the contested legitimacy of the weak Colombian state. From the nineteenth century it has been attempting to assert control over all of its territory and has faced a series of failed peasant revolutions for the past 150 years (Richani 2002: 23). At each stage of its development, the state has attempted to reform itself so as to protect the interests of an expanding and increasingly urban middle class. Right-wing paramilitaries (often linked with large landowners) provided the state with a private (but poorly controlled) security force while left-wing guerrilla groups sought to exploit the grievances of the dispossessed. In 2013, Oxfam reported that 80 percent of land in Colombia was held by 14 per cent of owners: an inequality that had a particular impact on women (Oxfam 2013). The weak state has been prone to regional influences from leftist ideologies, interventions from the United States and a ready supply of arms through porous borders. So, are drugs the cause of the Colombian conflict? No, but drugs money has become a fuel for a pre-existing conflict with long-term roots.

Sources: Richani (2002), Guáqueta (2007).

a slide into violent conflict. A group may become convinced of its own 'relative deprivation', especially if inequality is visible along religious, ethnic or racial lines (Jacoby 2008: 103–13). But, in such explanations, economic and development-related factors are secondary and only come into play when stimulated by other factors or if a prior existing condition (entrenched ethnic or religious difference) is in place.

The important point to bear in mind is that conflicts are caused by a combination of factors. Those who promote economic or development-related explanations for the outbreak of violent conflicts, need to take account of non-development-related explanations and how these interact with development-related factors. The philosopher and economist Amartya Sen makes the point that poverty on its own is not enough to cause conflict. He recalls his own childhood memories from Calcutta during the 1943 Bengal famine and 'the sight of starving people dying in front of sweetshop windows with various layers of luscious food displayed behind glass windows, without a single glass being broken, or law and order being disrupted' (Sen 2006: 143). Other factors, especially 'the illusion of singular identity ... in a world so obviously full of plural affiliations', were required to transform inequality and destitution into violent conflict (Sen 2006: 175).

Conflicts do not just happen. Just because a society is ethnically, racially or religiously fissured does not mean that conflict will follow. Seattle (a diverse multicultural city) is a more common model than Sarajevo (one with a history of ethnonational conflict). Indeed, given the multiplicity of identity groups that claim to be distinct from others, there is remarkably little violent conflict on the planet, and there is some evidence that it has declined over time (Brubaker and Laitin 1998; Pinker 2012). This suggests two points. The first is that many human societies have developed everyday peace systems that manage or suppress difference, often in non-violent ways (Mac Ginty 2014). The second is that violent conflict requires active instigation agents, particularly if latent tensions are to be escalated into overt violence. These instigation agents may take the form of political or community leaders who purposively inflame and mobilize their supporters, or circumstances – such as an assassination or shock election result – that agitate already tense group sensibilities (Zartman 2005: 268–73). Ukraine, Syria, Yemen, Nigeria, Lebanon and many other societies have experienced extended periods of calm (though not peace in a holistic sense) only for civil war to develop. This required active steps and reactions by political, military and community leaders.

Economics and civil war

Unsurprisingly, economists have been at the forefront of arguments that conflict causation can be explained by economics. Paul Collier and a number of collaborators have produced a corpus of studies that link the onset of civil war to economic factors (Collier and Hoeffler 2002; Collier et al. 2003). Many of these studies are based on econometric modelling and are attractive because they allow analysts to avoid considering nebulous and difficult-to-define factors such as identity or historical grievances. Two main arguments have been

Plate 3 *A Muslim cemetery in Bosnia: just because a society is ethnically fissured does not mean that conflict will follow.*

advanced under what has been termed the 'greed thesis' or economic explanations of violent conflict:

- That economic factors can act as predictors of violent conflict (or help identify civil war-prone societies)
- That combatants are motivated by economic predation.

The first argument identified economic factors – usually at the national level – that make a society prone to civil war. In particular, the level of per capita income, its rate of growth and the structure of the economy (especially its dependence on commodity exports) were identified as the key risk factors for the onset of civil war (on a dataset of 52 civil wars in the 1960–99 period). Collier and colleagues found that a doubling of per capita income halved the risk of civil war, and that when commodity exports account for 25 per cent of Gross Domestic Product (GDP), the risk of civil war is 33 per cent, as opposed to an 11 per cent risk of civil war if commodity exports are at 10 per cent (Bannon and Collier 2003: 2–3). Findings such as these have encouraged governments and policymakers to promote poverty reduction and economic diversification programmes – often based on free market remedies – as part of conflict prevention strategies (Brauer and Dunne 2012).

The second argument identified the profit motive – or 'greed' – as the primary motor behind civil war. Collier noted that 'conflicts are far more likely to be caused by economic opportunities than by grievance', but that rebel organizations will engage in a public discourse of grievance 'since they are unlikely to be so naive so as to admit to greed as a motive' (Collier 2000a: 91–2). Thus, 'civil wars occur where rebel organizations are financially viable', with the ability of antagonists to generate revenue being the principal reason why civil wars break out in some locations and not in others (Collier 2000b: 2). Münkler (2005) reinforces the view that civil war is economic rationalism taken to the extreme: the availability of weapons and untrained young men makes contemporary civil war 'downright cheap' and 'highly lucrative'. 'In the short term the force used in them yields more than it costs and the long-term costs are borne by others' (Münkler 2005: 74, 77). In this view, civil war conforms to a straightforward business model that seeks to maximize resource extraction through banditry, 'taxation' or the trafficking of diamonds, timber or people. It also aims to reduce costs by overlooking social responsibilities to citizens and cutting the costs of running a regular army. By boosting income and cutting costs, profit will be maximized. 'The entrepreneurs of the new wars' often emerged from the criminal underworld and used nationalist or ethnic movements as convenient vehicles from which to pursue their business interests (Münkler 2005: 80). William Reno's analyses of the civil wars in Sierra Leone and Liberia paint a dystopian picture in which political leaders drop all pretence of maintaining a functioning state that offers basic public services and protection to citizens (Reno 1997a, 1997b). Instead, leaders formed alliances with business organizations (often from overseas) to extract mineral resources. Ultimately, the civil wars in both territories resembled privatized conflict, with control of mineral resources the key aim (Keen 1998).

The claim by Collier and others that economic motivations must be given precedence over non-economic issues (such as identity) amounted to an intellectual drone strike aimed at scholars who held that ethnicity or religion held the key to the onset of civil war. They responded in kind, and were particularly annoyed that the economic rationalism arguments found favour with the world's main interna-tional financial institutions (IFIs), which were playing an increasing role in the management of conflict and post-war reconstruction. The econometric methodologies favoured by the greed theorists matched the bias of the IFIs towards rational quantifiable explanations for social phenomena and the technocratic, free market remedies they favoured. Critics of the greed thesis pointed out that societies experiencing civil war were a poor environment for the collection of statistics, and so urged caution over the datasets employed to suggest that certain countries offered a permissive economic context for the onset of civil war (Cramer 2002; Bensted 2011; Newman 2014: 22–3). Collier was also criticized for using proxies, using high male unemployment as a proxy for greed rather than grievance (Kandeh 2005: 96). But those who rejected the greed thesis or the economic explanations for civil war had two more serious objections. The first was that the greed thesis located the causes of war inside states and conveniently absolved external (mainly Western) actors from any blame (Demmers 2012: 112). This was especially the case in relation to the iniquitous international trading regimes that condemned many developing world states to prolonged economic retardation. The single state lens also tended to ignore regional dynamics, such as the flow of weapons and people across a border or interference from a neighbouring state, which often contributed to violent conflict. The second objection to the greed thesis was that its proponents mistook correlation for causation (Nathan 2005; Spear 2012: 232). Few denied that a permissive economic environment could encourage conflict or that a self-sustaining political economy of war could develop. What they did object to was the argument that economic factors were the *primary* engine of war. Instead, they argued that political and identity factors were the key initiation agents of war and that economic factors often subsequently came into play to change the nature and aim of the conflict.

Over time, the greed *or* grievance academic debate has given way to a consensus that greed *and* grievance are responsible for the outbreak of civil war (Ballentine and Nitzschke 2003; Murshed and Tadjoeddin 2009). The precise weight to be afforded to each is still contested, though since this weighting will change from conflict to conflict it is sensible to avoid building a general theory of conflict. We can say that most civil wars take place in poor countries, though poverty and inequality *per se* are not sufficient factors in the outbreak of civil war. Moreover, certain economic characteristics (such as low growth and a dependency on commodity exports) predispose societies to civil war. But a permissive environment does not amount to a causation factor.

Certainly economic factors can enable civil war, but for combustion to occur, the economic factors need to spark with other factors (Homer-Dixon 1994).

It is also important to note that development, rather than offering a ladder out of conflict, can contribute to conflict. Many of the social processes associated with development create conditions in which conflict is less easily restrained or is more easily escalated. As Box 1.2 and the example of cattle-raiding in Kenya and Tanzania show, urbanization, environmental degradation, the breakdown of family structures and a lessening of respect for traditional sources of dispute resolution may all create conditions permissive for violent conflict. Yet, to some, these social processes may simply be the by-products of social progress. China's rapid economic development illustrates the potential of development to contribute to conflict. An aggressive state-led development programme has resulted in the displacement of millions of people (well over a million people were displaced as part of the Yangtze Dam project: BBC News 2012), severe environmental degradation, land confiscations, and the perception among many rural peasants that they are

---

Box 1.2

Cattle-raiding in Kenya and Tanzania: development escalating conflict

Pastoral communities in Kenya and Tanzania have a long history of inter-tribal cattle-raiding (Fleischer 1998). Traditionally, the cattle-raiding was sustainable in that relatively small numbers of cattle were taken and casualty figures among the raiders and herders were low because traditional weapons were used. Often cattle raids were linked with rites-of-passage ceremonies, with adolescents using the raids as an opportunity to prove their valour (Kenya Human Rights Commission 2010). In recent years, however, development-related changes in society have transformed the character of cattle-raiding. As a result, the fallout of cattle-raiding – in terms of casualties and displacement – has increased markedly. There has been an increasing monetization of exchange, with the result that cattle-raiders are stealing cattle to sell to urban-based criminal gangs rather than for the traditional reasons of individual/group esteem and subsistence pastoral farming. Cattle raiders killed at least 32 Kenyan police officers in one ambush in 2012 (Macharia 2012). An increasingly urban and aspirational population is fuelling a demand for a meat-based diet, a demand that entrepreneurs are keen to satisfy. In addition to these development-related drivers, the intensity and effect of cattle-raiding have escalated as traditional weapons are being replaced by firearms (readily available from conflicts in the region).

the collateral damage in the country's economic liberalization. Significant levels of social unrest have been reported in China, a trend that has continued as its stellar growth rates have slowed (Göbel & Ong 2012: 8). The key point is that development often produces dislocation and uncertainty, and may materially disadvantage some groups, encouraging them to view their status in relation to other groups. In such circumstances the restraints on conflict may lessen.

The political economy of conflict maintenance

In some cases, the economic rationale of long-running violent conflicts is so apparent that it is possible to think that armed groups are motivated only by profit. Economic rationales often become more visible once a conflict is established and once markets and entrepreneurs have determined ways in which to exploit the opportunities of war. Here we might think of people smugglers in Syria, the illicit trafficking of oil out of Libya, and rebel groups that run mines in the Democratic Republic of Congo. But we can also think of the arms manufacturers who have an interest in states and armed groups using their weapons so that they need new supplies. We can also think of large corporations like G4S that run prisons in Israel that hold Palestinians (Neate 2014). When established, conflict economies can become self-perpetuating and entrepreneur-combatants may see few incentives to explore an end to the conflict. Economists have developed sophisticated models to show how looting and other forms of economic predation provide a powerful motive for combatants. But such models reveal little of the human character of civil war economies and the trials faced by citizens in time of war. Anthropologists, on the other hand, have succeeded in illustrating the extraordinary lengths to which individuals, families and communities go to in order to survive (Walker 2013; Bøås 2015). These studies show the adaptability of humans in their attempts to 'get by' and the extraordinary complexity of civil war economies (McIlwaine and Moser 2004).

A common misconception is that civil war economies are very much removed from the faraway mainstream economies of the developed world. The shiny shopping malls and online banking systems of the post-industrialized West seem a million miles away from civil war economies in which many economic transactions are illegal, unregulated or conducted under duress. But as anthropologist Carolyn Nordstrom (2008) shows, combatants and civilians in the midst of civil wars are often closely connected with the globalized international economy. This insight is important as it suggests that Western states and financial institutions – and indeed Western consumers – are complicit in the perpetuation of civil war economies in the developing world. Nordstrom (2008) argues that all civil wars rely on technologies (arms, communication, money transfers) and networks (trading partners and political supporters) that are to some extent international and transnational. As a result, the shadow economies of the civil war environment come into contact with the licit international economy: 'illicit profiteering must make use of legal production, transport, and monetary institutions' (Nordstrom 2008: 290). As David Keen (1998: 42) observes, 'even bandits need to sell what they steal.' The globalized 'buy/sell now, ask questions later' free market makes it easier for the licit and illicit economies to interact, and the sheer complexity of international markets (with multiple brokers) means that the paper trail from manufacturer/grower to consumer is easily obscured.

Nordstrom makes the case that rather than being marginal to the world economy, the apparently 'illicit' and 'shadowy' civil war economies are in fact central to it. Warlords and conflict entrepreneurs convert their illegally made profits into legal investments in the formal international economy or demand the same consumer goods that Western shoppers aspire to. Vast sums of money of dubious origin lubricate the international financial markets. Thus, drugs barons' money from Afghanistan and Colombia, once suitably laundered, is invested alongside the pension funds of Anglican bishops. In 2009, the UN estimated that transnational crime generated $870 billion annually, or about six times official development assistance (UNODC 2012). On top of this is the vast worldwide underground economy of dodging taxes and regulations. Just as consumers in war-torn societies demand goods and services from developed economies, consumers in the developed world demand goods from the sites of civil war. The high street shops and websites selling mobile phones (in 2014 93 per cent of adults in the United Kingdom owned a mobile phone (Ofcom 2015)) are just one end of a network of economic exchange stretching from the coltan mines of the Democratic Republic of Congo (DRC) from where an essential component in phone circuitry is extracted (Bøås 2014). Similarly, despite extensive international regulatory mechanisms, Sierra Leone's 'blood diamonds' manage to reach apparently legitimate retail outlets.

Plate 4 *A Porsche showroom in Beirut: people in war zones want the same consumer goods as those in peaceful countries.*

Corruption

Western analysts, and particularly the Western news media, can be shrill in their condemnation of corruption in developing world and civil war contexts. Certainly kleptocracy by ruling cliques and routine skimming by state functionaries can reach staggering proportions. Mohammed Soharto, Ferdinand Marcos and Mobutu Sese Seko are reputed to have embezzled a collective $50 billion during their respective reigns in Indonesia, the Philippines and Zaire (Denny 2004). The WikiLeaks US diplomatic cables contained the rumour that Sudanese President Omar al-Bashir squirrelled away $9 billion in overseas bank accounts (Hirsch 2010). But there is a big difference between the elite level siphoning off large sums of money, and the more common use of the underground economy by individuals and families who are simply trying to get by. The peculiar economic context of societies experiencing civil war means that we need to reassess our understanding of 'corruption' and 'illegal' market activities. Can corruption be said to exist if the formal economy has broken down and people need to rely on informal market mechanisms simply to survive?

The formal economy may be so dysfunctional (and often over-priced) that citizens have no choice but to operate in the informal sector. In cases of state collapse, there may be no legal economy at all. More commonly the state is weak and only able to regulate a fraction of economic activity within its borders. Estimates of the size of the informal economy in Africa range from 50 to 80 per cent of GDP (Benjamin 2014). The post-Taliban government in Afghanistan made massive strides in introducing a tax system. By 2013 it was able to raise $2.5 billion through tax revenues, customs duties and mining, but this was less than a half of the $7 billion needed for government expenditure. One news report summed up the views of many Afghans on the notion of paying tax: '...unlike developed countries where personal income tax generates a sizeable chunk of revenue, most Afghans scoff at the idea of giving the government some of their meager earnings. "It's not a good government," said moneychanger Abdurrahman Arif, 28, as he held a wad of soiled notes and scanned for customers. "I don't pay tax. The rich people don't and the government should go to them before they come to me"' (Houreld 2013).

Just as the formal, monetized and regulated economy is a way of life for most people in many Western states, the informal economy is a socially embedded behavioural and entirely rational norm in many civil war and post-civil-war societies. It makes sense to use non-patented medicines when patented medicines are either unavailable or exorbitantly priced. The formalization of the medical industry, through the protection of pharmaceutical patents, would not be in the interests of the vast majority of citizens because it would entail rocketing prices. The careless branding of certain economic activities as 'corrupt' or 'illegal' says as much about the Western worldview as it does about the activities themselves (Brown et al. 2004; Cheng and Zaum 2011). This is not to deny that corruption takes place and that it poses a real hazard to development and donor activity. Instead, it is to caution against the unthinking extension of Western yardsticks to non-Western, war-torn contexts. Box 1.3 illustrates the absurdity of some Western norms in war-affected societies. Kolstad et al. (2008) stress the importance of distinguishing between different types and scales of corruption. There is a difference between administrative informality and petty corruption found in everyday exchanges in a remote town where state officials feel the need to augment their wages, and large-scale frauds perpetrated by political leaders or senior bureaucrats. Simplistic moral and ethical judgements may not always be sensitive to

Box 1.3

A clash of cultures: Taliban-run Afghanistan and the British insurance industry

A British colleague, who worked for a major aid agency in Afghanistan in the mid-1990s, tells of his luggage being stolen on his arrival at Kabul airport. The luggage contained an expensive camera, so he thought it would be worth claiming on his worldwide travel insurance. He rang his UK-based insurer, who told him that they would post a claim form out to him (the Internet was in its infancy) and that he would need to get it stamped by the police in Kabul. He explained that the Taliban's police did not operate according to Western models of criminal justice and public safety, and that with no insurance industry operating in war-torn Afghanistan, they would have no knowledge of what the funny foreigner would be asking for. And anyway, the international postal service to Afghanistan was extremely unreliable. It was beyond the comprehension of the British insurer that a society would not have a police force like that in the UK and that they could not assist in the certification of insurance claims. The claim never got off the ground. The key point is that institutions and activities that may seem 'normal' in a Western environment do not necessarily have universal application. Effective bureaucracy and regulated markets, accepted components in Western states, may be uncommon in non-Western contexts.

the context in which corruption takes place. We should also be alert to the potential for international assistance to use and reinforce existing clientelistic and patronage networks. What begins as the rational use of 'local systems of disbursement' can reinvigorate networks that are less than transparent and may even reinforce warlord politics.

In some cases, international connections have served to prop up weak states and their corrupt patronage networks. Bayart (2000) notes how many African leaders have become adept at fobbing off international donors with the message they want to hear:

> [D]emocracy, or more precisely the discourse of democracy, is no more than yet another source of economic rents, comparable to earlier discourses such as the denunciation of communism or of imperialism in the time of the Cold War, but better adapted to the spirit of the age. It is, as it were, a form of pidgin language that various native princes use in their communication with Western sovereigns and financiers. Senegal, one of the main recipients of public development aid in sub-Saharan Africa, is a past master in this game of make-believe. It is no

> exaggeration to say that the export of its institutional image ... has
> replaced the export of groundnuts.
>
> (Bayart 2000: 226)

Essentially, the argument runs, international support has allowed corrupt regimes to continue systems of neo-patrimonialism and defer redistributive political and economic reform.

Legitimate monies and resources often reach civil war societies in the form of external donor aid. Once there, these funds risk fuelling the conflict. In an ethnically divided society, the infusion of external assistance is likely to be jealously scrutinized by all sides to make sure that their group gets 'its share'. Lebanon provides a complex example. The country played host to over one million Syrian refugees fleeing the civil war. Many of the refugees received basic care, including health care, from a range of UN bodies, bilateral donors and INGOs. But many Lebanese looked on with a deepening sense of jealousy. The Lebanese health care system is largely privatized and so some of the care available to refugees is not available to those in the host country (Amnesty International 2014). This resentment felt by the host community was refracted through the sectarian lens of an already divided Lebanon causing further tensions.

External humanitarian and development agencies are often placed in an invidious position in war-torn societies. Do they sit on the sidelines, refusing to give assistance for fear that aid may fall into the 'wrong hands', or do they muck in and hope that their efforts help the genuinely needy despite the risks? Médecins Sans Frontières (MSF) staff in Myanmar/Burma faced this problem during the years of the military dictatorship. The organization wanted to treat malaria and AIDS sufferers in extremely poor areas but needed military permission to do so. Senior staff were faced with the dilemma of 'playing golf with the generals' (a way of talking to them informally) or sitting on their hands. The danger of dealing with the generals was that it might confer legitimacy onto military thugs (Terry 2011: 109–11). Civil war contexts are unlikely to leave ethical principles unscathed, yet that means material and symbolic resources being at the mercy of combatants. Moreover, in some contexts, the sheer scale of donor assistance can distort the economy. As one observer noted, 'other than the state itself, the aid business is today the single biggest employer in most African states' (van de Walle 2001: 58).

Just as there is a political economy of war, there is a political economy of humanitarianism and development assistance. Perhaps this is most visible in the micro-economies that spring up to service international humanitarian workers in war-affected societies. The cluster of Western-style bars and fast food outlets are often identifiable because of the white 4×4s parked outside. But at the macro-economic level, as will be discussed in later chapters, international economic and development interventions are often deliberately aimed at reshaping the war-affected economy to reconnect it with the formal global economy. Many of the liberal economic 'reforms' actually lead to the further immiseration of citizens: state employees are sacked ('rightsizing bureaucracy'), debts run up by previous regimes must be paid ('respecting international financial obligations'), prices rise as exchanges are formalized and monetized ('regulation'), and indigenous businesses cannot compete with cheap imports ('the global free market'). Kiely (2007: 434) notes how 'liberalisation undermines the capacity of developing countries to develop dynamic comparative advantages', yet liberalization seems to be the main tool in the international toolbox.

The chief points of this section are that as war further distorts econo-mies, people in war-affected societies (combatants and civilians alike) will take extraordinary measures to survive and this may involve activities that Western observers may judge 'corrupt' or 'illegal'. But outside observers may be hypocritical in such judgements: civil war economies are hardwired into the very fabric of the formal interna-tional economy. Consumer demand in Western states and the interna-tional economic structures erected by Western states influence the choices and constraints faced by people on the ground in civil war societies. While we can paint an abstract picture of the economic impact of civil war (the 'typical' civil war costs $50 billion (Collier 2004)), it is important that we recognize the human experience of civil war and how many people are brutalized and humiliated by civil war economics, whether by being trafficked, being forced to sell family heirlooms or living in an environment in which theft is regarded as a normal survival mechanism (Mac Ginty 2004).

Natural resources and conflict

As already noted, economists have claimed that an economic dependency on natural resource exports increases the likelihood of the outbreak of civil war. But the mere presence of natural resources does not lead to armed conflict. As Cramer (2006: 117) observes, 'scarcity and violence are a product of social relations rather than inherent in the relative abundance of a particular good, object or resource.' In other words, it is the management of the resources that really matters. It is notable that Norway, with its abundant natural gas supplies, high taxes, and social provision, tops most global living standards tables. Yet, despite being 'blessed' by nature's largesse, a significant number of resource-rich states are chronically poor. Prominent in the list of mineral-dependent countries (reliant on extractive minerals for 25 per cent of non-tangible exports) were the Democratic Republic of Congo, Somalia and Libya – all the sites of conflict (Hagland 2011). It is the labour practices (voluntary or coerced), distribution of licences (open competition or patronage) and destination of profits (public or private coffers) that will determine whether states face a resource curse or windfall. Patterns of land ownership are particularly important in developing world contexts, as access to land (and the quality of that land) may afford subsistence and thus some measure of autonomy in economic matters (Miall 2007: 125–9).

Crucially, the perception of the management of resources is important. In a number of cases, minority ethnic, nationalist or religious groups have pursued grievances stemming from allegations that the state was plundering 'their' natural resources with few obvious benefits in return. Thus the Acehnese in Indonesia, Muslims in Mindanao (Philippines), Christians in southern Sudan and the Ogoni in the Niger Delta (see Box 1.4) have all engaged in secessionist conflict with the state and have campaigned for their 'fair share' of returns from resource exploitation. Indeed, a recurring narrative in debates for Scottish independence concerns who has the right to the receipts from North Sea gas: the Scottish or the whole of the United Kingdom? Although natural resources play a crucial role in these conflicts, it is incorrect to conceive of them as pure 'resource wars'. Instead, the conflict arises from a complex mix of the presence of resources, the pattern of resource exploitation, the perception of the benefits of that exploitation and identity affiliations. If the stakes are high, identity affiliations can mutate, with groups and individuals attaching increasing weight to the purity of their ethnic group and rediscovering (or inventing) their 'unique' history (Wilmer 2002: ix). In such ways, conflicts become 'ethnicized' and exclusion from the benefits of natural resources may provide a powerful impetus to escalate conflict.

Box 1.4

Oil extraction in Nigeria's Niger Delta

Nigeria's oil-rich Niger Delta region has experienced significant levels of conflict, criminal violence and environmental damage for several decades. The central govern-ment has encouraged major oil corporations to exploit the oil reserves, but residents in the Delta region claim that any benefits bypass local communities and that pollution has seriously affected quality of life. Armed criminal gangs regularly steal oil from pipelines and kidnap foreign oil workers for ransom, while other local groups have been vocal in their condemnation of the profiteering, corruption and environmental dis-regard of 'imperial Abuja' or have encouraged labour unrest. On top of this, there have been clashes between rival ethnic groups. Oil company attempts to co-opt tribal chiefs through payments have made chieftaincies extremely lucrative, sparking conflict and changing community perceptions of chiefs. In response to the oil-related tension and violence, the Nigerian government has variously declared a state of emergency and sent in the army, attempted to buy off local leaders, reassured foreign oil companies, promised to invest a greater share of oil in the region and occasionally mounted prose-cutions against corrupt officials (including one against two rear-admirals accused of stealing an oil tanker: Clayton 2005). The violence has seriously disrupted oil produc-tion, but the potential rewards for the government, local and national political leaders, and overseas oil companies are simply too great for anyone to contemplate withdraw-ing from the region (Obi and Rustad 2011).

Sources: Clayton (2005), Obi and Rustad (2011).

Winston Churchill's observation that 'God put the West's oil under Middle Eastern feet' is a reminder of competitive geo-strategic interests in natural resources that make resource-rich developing world states prone to intervention by powerful states. Quite simply, advanced and developing economies are dependent on oil. Their ways of life, politics and economics would be utterly unsustainable if ready access to oil were not secured. In the main, the market has been successful in procuring oil, and major Western states are not as dependent on oil from conflict-affected areas as some analysts sug-gest. The United States was the world's third largest oil producer in 2012 (EIA 2013), and most oil-producing Gulf states are compliant with Western economic and geopolitical strategy. Indeed, the oil profits of most Gulf states are tied up in the New York and London stock markets, so oil-rich states have no incentive to spark economic instability by coming together and attempting to use leverage over oil supply. But as demand for oil in the developing world (especially

India and China) surges, pressure on apparently finite oil resources increases. Both the market and states can be expected to act in self-interested ways to ensure continued access to energy. As the cases of post-Saddam Iraq, Colombia and Nigeria show, consumer demand is so great that complexes of private companies and state bodies will come together to create oases of petrochemical calm in the midst of wider conflict just to ensure that oil continues to flow.

Demand for oil, water, timber, diamonds and other minerals continues to be the source – or at least fuel (literally in some cases) – of violent conflict. As populations become richer, the natural resources of landscape and aesthetic beauty come under increasing pressure from tourism. This is despite a growing awareness of the environmental costs of most forms of tourism. In a number of cases, Western 'tourist bubbles' exist alongside local inequality, repression or conflict (Rogers 2000: 2). The resorts of the Maldives have soared in popularity, despite the suppression of political opposition by the ruling regime and the floggings for 'fornication' (Amnesty International 2015: 241–2). Perhaps the starkest juxtaposition of the Western tourism industry with conflict has been the Royal Caribbean Cruise Line's virtual annexation of 260 acres of Haiti as a stopping-off point for its cruise liners. Branded as 'Fantasy Island', 'Magic Island' and 'Labadee®', the term 'Haiti' is barely mentioned and no mention is made of Haiti's history of civil war or its 2010 earthquake that killed about 160,000 people and displaced over one million (Walker 2010; Royal Caribbean 2015). Passports are not stamped as tourists enter the leisure enclave and visitors are warned not to venture beyond the resort boundaries (Orenstein 1997). But over those walls in Haiti proper Gross National Income per capita was US$810 per year in 2013 (World Bank 2015). Labadee® represents one version of the 'no war, no peace' phenomenon, whereby violent conflict and poverty is compartmentalized in certain parts of the state, allowing other parts to function as 'normal' (Mac Ginty 2006).

Conclusions

There is growing consensus among policymakers, academics and others that the escalation, maintenance and transformation of conflict are linked to development. The precise nature of these linkages is still debated. In truth, there is no exact science linking conflict to development; the constellation of variables is simply too great. Despite the apparent sophistication of econometric modelling or the 'seen it all' world-weary cynicism of development practitioners, conflict interventions and development programming contain a good deal of guesswork and finger-crossing. Western donor governments, such as the Department for International Development (DFID) in the UK, show a commitment to conflict-sensitive programming, or development interventions calibrated to have a minimal or positive impact on a conflict situation. But the volatility and lack of regulation in many war-torn or post-conflict societies mean that the precise effect of development inputs is unknown. The bottom line is that development or reconstruction assistance is a resource, and political and militant actors will act rationally in attempting to maximize their access to, or benefit from, those resources. Despite good intentions, development inputs may actually fuel conflict. Moreover, and as will be discussed in Chapter 2, many of the factors that influence development in a war-torn society will be external to that society. Just as many citizens may feel powerless in the midst of conflict, they may also feel that development is a process that is *done to* them.

The relationship between security and development is fraught with thorny questions, especially in the post-9/11 world in which the 'security imperative' is easier to justify among many audiences than a 'development imperative'. When Islamic State (IS) occupied large parts of Iraq from 2013 onwards, the Western response was one of bombing from the air, not of examining development alternatives that could defeat IS through livelihoods and economic advancement among the populace. Development and reconstruction require a certain level of order. At its most benign, this order can take the form of institutionalization and regulation; in effect a process of 'normalization' whereby the uncertainty of a violent context is replaced by the certainty of stability. But in a less benign scenario, there can be unacceptable costs associated with 'stabilization', such as restraining civil liberties or empowering private security contractors. In effect, a process of development and post-war reconstruction can be as disempowering as the war once was. If a holistic view of development is taken, in which development extends far beyond the narrow confines of economic growth, then the process of development might also involve de-development.

Summary

- Explanations of the causes of violent conflict have focused on issues of 'greed' (economic causes) and 'grievance' (such as identity).
- Most scholars argue that a mixture of greed and grievance factors contribute to violent conflict.
- The factors that cause a violent conflict may differ from the factors that sustain a conflict.
- Resources alone, such as oil, diamonds or water, do not cause conflict. What is important is the nature of the extraction and management of those resources.
- Just as violent conflict can distort an economy, so too can aid and peace-support, with issues of 'corruption' gaining increasing attention in recent years.

Discussion questions

- Do you find that the criticisms of the greed thesis of conflict causation are justified?
- What are the economic factors that can sustain violent conflict once it has started?
- Can there be corruption if the formal economy has broken down?
- Should aid agencies halt all assistance to a conflict area if they know that some of their aid will be siphoned off by combatants, or should they take this as a necessary evil of operating in a conflict zone?

Further reading

There is an enormous literature on conflict causation. A good starting point for the econometric perspective is P. Collier et al. (2003) *Breaking the Conflict Trap: Civil war and development policy*, Washington, DC: World Bank and Oxford University Press, and P. Collier (2007) *The Bottom Billion: Why the poorest countries are failing and what can be done about it*, Oxford: Oxford University Press. Wider studies of conflict that make connections between development and conflict include works by Mark Duffield, Paul Jackson, Eleanor O'Gorman, Mary Kaldor, David Keen and Christopher Cramer. A particularly useful way of following developments in conflict zones is to search for blogs by people living there or by aid workers. A few years after the conflict, autobiographies might emerge as well that give insights into living amidst conflict.

Useful websites

The World Bank has many documents on Fragility, Conflict and Violence: http://www.worldbank.org/en/topic/fragilityconflictviolence. The Uppsala Conflict Data Program contains probably the most comprehensive source of conflict statistics (http://www.pcr.uu.se/research/UCDP/), while the Global Peace Index (http://www.visionofhumanity.org/#/page/our-vision) collects country-wide information on development and peace. Useful information can also be found via the International Crisis Group (www.crisisgroup.org), Peace Direct (peacedirect.org) and International Alert (http://www.international-alert.org). On resource exploitation, details of attempts to reduce the trade in blood diamonds can be found at www.kimberleyprocess.com/. Details of the anti-corruption organization Tiri can be found at www.tiri.com.

References

Amnesty International (2014) Syrian refugees in Lebanon desperate for health care amid international apathy. Amnesty International, 21 May. Available at: https://www.amnesty.org/en/articles/news/2014/05/syrian-refugees-lebanon-desperate-health-care-amid-international-apathy/ (accessed on 8 September 2015).

Amnesty International (2015) *Amnesty International Report 2014/2015: The state of the world's human rights*. London: Amnesty International. Available at: https://www.amnesty.org/en/documents/pol10/0001/2015/en/ (accessed on 20 April 2015).

Appleby, R.S. (2000) *The Ambivalence of the Sacred: Religion, violence and reconciliation*. Lanham, MD: Rowman and Littlefield.

Ballentine, K. and Nitzschke, H. (2003) *Beyond Greed and Grievance: Policy lessons from studies in the political economy of armed conflict*. New York: IPA Policy Report.

Bannon, I. and Collier, P. (2003) Natural Resources and Conflict: What Can We Do. In Bannon, I. and Collier, P. (eds) *Natural Resources and Violent Conflict*. Washington, DC: World Bank, pp. 1–16.

Bayart, J.F. (2000) Africa in the World: a history of extraversion. *African Affairs* 99: 217–67.

BBC News (2012) China's Three Gorges Dam may displace another 100,000. 18 April. Available at: http://www.bbc.co.uk/news/world-asia-china-17754256 (accessed on 8 September 2015).

Benjamin, N (2014) Informal economy and the World Bank. World Bank Research Working Paper no. 6888, May.

Bensted, R. (2011) A critique of Paul Collier's 'greed versus grievance' thesis of civil war. *African Security Review* 20(3): 84–90.

Bøås, M. (2015) *The Politics of Conflict Economies: Miners, merchants and warriors in the African borderland*. London: Routledge.

Brauer, J. and Dunne, J.P. (2012) *Peace Economics: A macroeconomic primer for violence affected states*. Washington, DC: USIP Press.

Brown, E., Cloke, J., Sohail, M. (2004) Key myths about corruption. Briefing paper for a workshop on corruption and development presented at the Development Studies Association Annual Conference, Westminster, London. 6 November.

Brown, M.E. (1997) The Causes of Ethnic Conflict: An Overview. In Brown, M.E., Coté, O.R., Lynne-Jones, S.M. and Miller, S.E. (eds) *Nationalism and Ethnic Conflict*. Cambridge, MA: MIT Press, pp. 3–25.

Brubaker, R. and Laitin, D.D. (1998) Ethnic and nationalist violence. *Annual Review of Sociology* 24: 423–54.

Cheng, C. and Zaum, D. (eds) (2011) *Post-Conflict Peacebuilding: Selling the peace?* London: Routledge.

Clayton, J. (2005) Nigerian admirals pay the price for stealing captured oil tanker. *Times Online*, 8 January. Available at: http://www.timesonline.co.uk/tol/news/world/article409606.ece (accessed on 13 June 2008).

Collier, P. (2000a) Doing Well out of War: An Economic Perspective. In Berdal, M. and Malone, D. (eds) *Greed and Grievance: Economic agendas in civil wars*. Boulder, CO: Lynne Rienner, pp. 91–111.

Collier, P. (2000b) *Economic Causes of Civil Conflict and their Implications for Policy*. Washington, DC: World Bank.

Collier, P. (2004) Development and Security. 12th Bradford Development Lecture, University of Bradford, 11 November.

Collier, P. and Hoeffler, A. (2002) *Greed and Grievance in Civil Wars*. Working Paper Series. Oxford: Centre for the Study of African Economies.

Collier, P., Elliot, V.L., Hegre, H., Hoeffler, A., Reynal-Querol, M., Sambanis, N. (2003) *Breaking the Conflict Trap: Civil war and development policy*. Washington, DC: World Bank and Oxford University Press.

Connor, W. (1994) *Ethnonationalism: The quest for understanding*. Princeton, NJ: Princeton University Press.

Cramer, C. (2002) Homo Economicus goes to war: methodological individualism, rational choice and the political economy of war. *World Development* 30(11): 1845–64.

Cramer, C. (2006) *Civil War is not a Stupid Thing: Accounting for violence in developing countries*. London: Hurst.

Demmers, J. (2012) *Theories of Violent Conflict: An introduction*. London: Routledge.

Denny, C. (2004) Suharto, Marcos and Mobutu head corruption table with $50bn scams. *Guardian*, 26 March.

EIA (2013) How dependent are we on foreign oil? US Energy Information Administration, 10 May. Available at: http://www.eia.gov/energy_in_brief/article/foreign_oil_dependence.cfm (accessed on 20 April 2015).

Fleischer, M. (1998) Cattle raiding and its correlates: the cultural–ecological consequences of market-oriented cattle raiding among the Kuria of Tanzania. *Human Ecology* 26(2): 547–72.

Göbel, C. and Ong, L.H. (2012) *Social Unrest in China*. London: Chatham House.

Guáqueta, A. (2007) The way back in: reintegrating illegal armed groups in Colombia then and now. *Conflict, Security and Development* 7(3): 417–56.

Hagland, D. (2011) *Blessing or Curse? The rise of mineral dependence among low and middle income countries*. Oxford Policy Management Group Report, December. Oxford: OPMG.

Hirsch, A. (2010) WikiLeaks Cables: Sudanese President 'stashed $9bn in UK banks'. *Guardian*, 17 December.

Homer-Dixon, T. (1994) Environmental scarcities and violent conflict: evidence from cases. *International Security* 19(1): 5–40.

Houreld, K. (2013) Afghans warned: The taxman is coming after you. Reuters, 31 March. Available at: http://www.reuters.com/article/2013/03/31/us-afghanistan-tax-idUSBRE92U00Q20130331 (accessed on 8 September 2015).

Huntington, S. (1998) *The Clash of Civilizations and the Remaking of World Order*. London: Touchstone.

Jacoby, T. (2008) *Understanding Conflict and Violence: Theoretical and interdisciplinary approaches*. London: Routledge.

Kandeh, J. (2005) The Criminalization of the RUF Insurgency in Sierra Leone. In Arnson, C. and Zartman, I.W. (eds) *Rethinking the Economics of War: The intersections of need, greed and creed*. Baltimore, MD: Johns Hopkins University Press, pp. 84–106.

Kaplan, R. (1994) The coming anarchy. *Atlantic Monthly*, February.

Keen, D. (1998) *The economic functions of violence in civil war*. Adelphi Paper 320. London: International Institute of Strategic Studies.

Kenya Human Rights Commission (2010) *Morans No More: The changing face of cattle rustling in Kenya*. Nairobi: Kenya Human Rights Commission.

Kiely, R. (2007) Poverty reduction through liberalisation? Neo-liberalism and the myth of global convergence. *Review of International Studies* 33: 415–34.

Kolstad, I., Fritz, V. and O'Neil, T. (2008) Corruption, anti-corruption efforts and aid: Do donors have the right approach? Dublin: Advisory Board for Irish Aid. Available at: http://www.odi.org.uk/PPPG/politics_and_governance/publications/GAPWP3.pdf (accessed on 18 June 2008).

Mac Ginty, R. (2004) Looting in the context of violent conflict: a conceptualization and typology. *Third World Quarterly* 25(5): 857–70.

Mac Ginty, R. (2006) *No War, No Peace: The rejuvenation of stalled peace processes and peace accords*. London: Palgrave.

Mac Ginty, R. (2014) Everyday peace: bottom-up and local agency in conflict-affected societies. *Security Dialogue* 45(6): 548–64.

Macharia, J. (2012) At least 32 police dead in cattle raid ambush. Reuters, 12 November. Available at http://uk.reuters.com/article/2012/11/12/us-kenya-security-idUSBRE8AB0NL20121112 (accessed on 8 September 2015).

McIlwaine, C. and Moser, C. (2004) *Encounters with Violence in Latin America: Urban poor perceptions from Colombia and Guatemala*. London: Routledge.

Mandelbaum, M. (2002) *The Ideas that Conquered the World: Peace, democracy and free markets in the twenty-first century*. New York: PublicAffairs.

Miall, H. (2007) *Emergent Conflict and Peaceful Change*. Cambridge: Polity.

Münkler, H. (2005) *The New Wars*. Cambridge: Polity.

Murshed, S.M. and Tadjoeddin, M.Z. (2009) Revisiting the greed and grievance explanations for violent internal conflict. *Journal of International Development* 21(1): 87–111.

Nathan, L. (2005) The frightful inadequacy of most statistics: A critique of Collier and Hoeffler on the causes of civil war. LSE Crisis States Development Research Centre, Discussion paper 11, September.

Neate, R. (2014) Desmond Tutu tells G4S to stop supplying Israeli prisons. *Guardian*, 4 June. Available at: http://www.theguardian.com/business/2014/jun/04/desmond-tutu-g4s-israeli-prisons-noam-chomsky (accessed on 8 September 2015).

Newman, E. (2014) *Understanding Civil Wars: Continuity and change in intrastate conflict*. London: Routledge.

Nordstrom, C. (2008) Casting Long Shadows: War, Peace and Extralegal Economies. In Darby, J. and Mac Ginty, R. (eds) *Contemporary Peacemaking: Conflict, peace processes and post-war reconstruction*. Basingstoke: Palgrave, pp. 289–99.

Obi, C. and Rustad, S.A. (eds) (2011) *Oil and Insurgency in the Niger Delta*. London: Zed.

Ofcom (2015) *Facts and figures*. Ofcom website. Available at: http://media.ofcom.org.uk/facts/ (accessed on 8 September 2015).

Orenstein, C. (1997) Fantasy Island: Royal Caribbean parcels off a piece of Haiti. *The Progressive*. Available at: http://findarticles.com/p/articles/mi_m1295/is_n8_v61/ai_19622661/pg_1?tag=artBody;col1 (accessed on 21 July 2008).

Pinker, S. (2012) *The Better Angels of our Nature: A history of violence and humanity*. London: Penguin.

Posen, B. (1993) The Security Dilemma and Ethnic Conflicts. In Brown, M. (ed.) *Ethnic Conflict and International Security*. Princeton, NJ: Princeton University Press, pp. 103–24.

Reno, W. (1997a) War, markets, and the reconfiguration of West Africa's weak states. *Comparative Politics* 29(4): 493–510.

Reno, W. (1997b) African weak states and commercial alliances. *African Affairs* 96: 165–85.

Richani, N. (2002) *Systems of Violence: The political economy of war and peace in Colombia*. New York: SUNY Press.

Rogers, P. (2000) *Losing Control: Global security in the twenty-first century*. London: Pluto.

Royal Caribbean (2015) Caribbean cruises – destinations. Royal Caribbean website. Available at: http://www.royalcaribbean.co.uk/destinations/caribbean/destinations.aspx (accessed on 20 April 2015).

Sen, A. (2006) *Identity and Violence*. New York: Norton.

Simmel, G. (1955) *Conflict: The web of group affiliations*. New York: Free Press.

Spear, J. (2012) Trade and Resources: A Security Perspective. In Spear, J.D. and Williams, P.D. (eds) *Security and Development in Global Politics: A critical comparison*. Washington, DC: Georgetown University Press, pp. 229–46.

Tajfel, H. (1978) Social Categorization, Social Identity and Social Comparison. In Tajfel, H. (ed.) *Differentiation between Social Groups: Studies in the social psychology of intergroup relations*. London: Academic Press.

Terry, F. (2011) Myanmar: Playing Golf with the Generals. In Magone, C., Neuman, M. and Weissman, F. (eds) *Humanitarian Negotiations Revealed: The MSF experience*. London: Hurst, pp. 109–26.

Tilly, C. (1985) War Making and State Making as Organized Crime. In Evans, P., Rueschemeyer, D. and Skocpol, T. (eds) *Bringing the State Back In*. Cambridge: Cambridge University Press, pp. 169–91.

UNODC (2012) New UNODC campaign highlights transnational organized crime as a US$870bn a year business. United Nations Office on Drugs and Crime, 16 July. Available at: http://www.unodc.org/unodc/en/frontpage/2012/July/new-unodc-campaign-highlights-transnational-organized-crime-as-an-us-870-billion-a-year-business.html (accessed on 8 September 2015).

Van de Walle, N. (2001) *African Economies and the Politics of Permanent Crisis, 1979–1999*. Cambridge: Cambridge University Press.

Walker, J. (2010) Royal Caribbean 'returns' to its trademarked private fantasy island of Labadee – while Haiti suffers. Jim Walker's Cruise Law News. Available at: http://www.cruiselawnews.com/2010/01/articles/caribbean-islands/royal-caribbean-returns-to-its-trademarked-private-fantasy-island-of-labadeea-while-haiti-suffers/ (accessed on 20 April 2015).

Walker, R. (2013) *Enduring Violence: Everyday life and conflict in eastern Sri Lanka*. Manchester: Manchester University Press.

Wilmer, F. (2002) *The Social Construction of Man, the State, and War: Identity, conflict, and violence in the Former Yugoslavia*. London: Routledge.

World Bank (2015) Haiti. Available at: http://data.worldbank.org/country/haiti (accessed on 8 September 2015).

Young, C. (2003) Explaining the Conflict Potential of Ethnicity. In Darby, J. and Mac Ginty, R. (eds) *Contemporary Peacemaking: Conflict, violence and peace processes*. Basingstoke: Palgrave Macmillan, pp. 9–18.

Zartman, I.W. (2005) Need, Greed and Creed in Intrastate War. In Arnson, C. and Zartman, I.W. (eds) *Rethinking the Economics of War: The intersection of need, greed and creed*. Baltimore, MD: Johns Hopkins University Press, pp. 256–84.

2 Institutions: hardware and software

The twentieth century is often termed 'the American century' and, thus far, the twenty-first century continues the trend of US dominance. India, the European Union, and particularly China, are growing in influence and economic might, but they have yet to eclipse the US. Importantly, they are shy of directly challenging the US and have strong vested interests in stability under US stewardship. As well as being dominated by large powers, the twentieth century, particularly from 1945 onwards, was a century of international institutions. Institutions such as the United Nations, NATO, African Union (AU) and EU continue to play a key role in thwarting and facilitating conflict and development in the twenty-first century. Indeed, a select group of international institutions comprises the primary international instruments dedicated to preventing and minimizing the impact of war, spearheading post-war reconstruction and promoting development. At the same time, a select group of international institutions is often blamed for underdevelopment and de-development, and by extension contributing to conflict. To complicate matters, the same institutions have been blamed for contributing both to the escalation and transformation of violent conflict.

This chapter sketches the principal international architecture that provides the context for contemporary conflict and development. Just as it is difficult to discuss computing hardware in isolation from software, it is difficult to gain a comprehensive understanding of international political structures without discussing their operating 'software' or their behaviour. The chapter begins by discussing software and then moves on to hardware. Under software, we discuss the liberal peace or the overarching philosophy that shapes many international peace-support operations, post-war reconstruction programmes and development interventions. International institutions are given shape and purpose by the principles, behaviour and world-view of their key members. It is argued here that the liberal peace has a peculiarly Western flavour that reinforces the dominance of existing elites and promotes highly specialized Western ideas, namely versions of liberalism, democracy and economics. As will be shown, the liberal peace has decidedly illiberal aspects. A crucial element of the liberal peace is hyper-globalization which supports multiple connections through multiple networks so that conflict and development on various parts of the planet are linked. The software section ends by considering alternatives to the liberal peace, and finds that spaces for these are often marginal.

The chapter then moves on to discuss 'hardware' or the principal institutions that comprise the international political system. It begins with a brief discussion of the key constituent feature of the international system: the state. The important factor from our point of view is the political organization of the state, particularly in terms of its relationships with its citizens and market, and its ability to resist or adapt to external pressure. The chapter then reviews the roles and effectiveness of the primary collective security and development promotion organizations (the UN, EU, etc.), before examining the role of the international financial institutions (the World Bank, IMF, etc.). The chapter concludes by recommending that we adopt the lens of complex multilateralism when reviewing the role of international institutions in relation to conflict and development. In other words, we need to move away from a view in which we have compartmentalized entities such as states and international organizations that have formal and well-defined linkages between them. Instead, conflict and development operate in a much more complex environment in which multiple transnational and international actors cooperate and clash. This 'cast list' is much more extensive than the traditional list of states and international institutions, and includes globalized multinational companies, NGOs and transnational social movements.

A final introductory point is to highlight the importance of resisting US-bashing as the catch-all explanation for the state of international relations, development and conflict. Yes, the US is the hegemon, or the world's most powerful state, and has global interests and almost global reach. Yet, rather than simply bashing the US, it is important to ask why it acts in the ways it does. It acts in usually rational ways to safeguard its own interests. Moreover, the US top-dog position suits many others too and they are active supporters of the stability and creed of economic growth that it supports.

Software

Without guidance, international institutions such as the United Nations, the Asian Development Bank or the African Union are merely empty vessels. They do not have autonomous lives of their own. Instead, they reflect the positions of their most powerful members and lobbies. Crucially, international organizations are not neutral (despite professions otherwise). International organizations are the product of a political, economic and cultural reality that has been, and is, heavily contested. At the heart of this struggle is the old-fashioned concept of 'power', a concept that once dominated political science and international relations and now tends to be overlooked. It is not over-simplistic to say that a basic struggle between power-holders and power-seekers defines the structure and operations of many international organizations. The power-holders, who are often rich Western states, international organizations, international financial institutions and corporate interests, wish to maintain their stranglehold over economic, political and cultural power over the power-seekers. The latter are often developing world and sometimes conflict-ridden states and their populations, minority communities, the dispossessed and those of a critical perspective.

The 'software' that directs international institutions to act in prescribed ways is often specifically designed to reinforce the position of power-holders and thwart attempts by power-seekers to achieve a more egalitarian share of resources. The power-holders, and particularly the hegemon in the form of the United States, have been remarkably successful in perpetuating their power. Tim Jacoby (2007: 523) notes how the hegemon is skilled at convincing states that its perpetual dominance is in their own interests, maintains enough military power to cow opposition, and guarantees its material superiority through a heavily biased distribution of resources and capital. This section discusses 'software' or the global computer program that helps run the dominant international political system and many of the international organizations within it.

The liberal peace

Different scholars use different lenses with which to interpret the world. Feminists, for example, might argue that the main software package that drives the international political system is the patriarchy or the male dominance that is infused into many aspects of life (Tickner 2001). Marxists might be tempted to interpret their known universe through an analysis of the means of production and patterns of ownership and consumption (Maclean 1988). This study finds that the liberal peace lens is particularly useful in explaining many of the meta-influences at work in situations of conflict and development. The liberal peace is a highly specialized form of peace and development intervention promoted by leading Western states, leading international organizations and the international financial institutions in their attempts to shape the international political system and its constituent parts (Paris 2004). It uses the language of liberalism (democracy, freedom, free markets, human rights), hence the phrase 'liberal peace'. Whether it is actually liberal in its execution is a matter of great debate (Campbell et al. 2011; Roberts 2011; Tadjbakhsh 2011; Richmond and Mac Ginty 2014). The liberal peace is the 'ideology upon which life, culture, society, prosperity and politics are assumed to rest' (Mac Ginty and Richmond 2007a: 493). It is capable of constructing an attractive rationale for its own promotion. Thus it speaks of 'responsibility', 'development', 'common interests' and above all, intervention (Williams 2007: 543).

Sometimes called 'liberal interventionism' or 'liberal internationalism', the liberal peace is most visible in societies undergoing Western-backed peace-support interventions in the aftermath of civil war. But many of the tools of the liberal peace, particularly in disciplining societies, governments and economies, are also at work in developing states that have not experienced recent war. In non-post-war environments, these interventions are often covered by the terms 'good governance' and 'reform' and we find the same commitment to the market as a prerequisite for debt relief and poverty reduction strategy funding (Craig and Porter 2003; Abrahamsen 2004). Indeed the 2013–15 disciplining of an indebted Greece by the European Commission, the European Central Bank and the International Monetary Fund is instructive of the power wielded by an elite group of states and institutions. Greece, a democratic sovereign country, and a member of NATO and the EU, was bullied into accepting loan extension deals that were unpopular and clearly disadvantageous to the Greek people (through cuts to health care, pension and wages). Creditors used the language of responsibility and living within one's means, but it was clear that liberal economic internationalism was backed up with coercion and threats. Greek Finance Minister Yanis

Plate 5 *The UN in Jordan: the UN can be considered as an agent of the liberal peace.*

Varoufakis gave a startling insider account of the talks trying to avert Greece's exit from the Eurozone: 'You had might as well sung the Swedish national anthem, you'd have got the same reply … You either sign on the dotted line or you are out' (cited in Lambert 2015).

An advantage of the liberal peace tool is that it allows us to make comparisons, sometimes across contexts that may not have obvious similarities or connections. It also allows us to make sense of the strategy employed by the leading political and economic actors in the international system. The liberal peace can be seen as a normatively neo-liberal system of compliance that is variously recommended, induced and enforced by leading states, leading international organizations and international financial institutions. Developing world states and states emerging from conflict often have little choice but to accept the liberal peace. If they do not conform, for example like Cuba, then they risk being locked out of access to funds and international platforms.

Many aspects of the liberal peace are deeply illiberal. It promotes a highly specialized form of liberalism that is often prescriptive and reflective of Western norms (Mac Ginty 2006: 33–57). Rajiv Chandrasekaran's (2007: 7) exposé of life inside Baghdad's Green Zone, in the aftermath of the 2003 invasion, provides a stark illustration of the crass ethnocentrism at the heart of the liberal peace as manifested in Iraq's Coalition Provisional Authority (CPA). He tells of how US administrators lived in a hermetically sealed compound complete with air-freighted fast food, US sports television channels and other home comforts. The security situation meant that many administrators rarely left the compound, while only a very few Iraqis could gain access to their new rulers. Many CPA-staffers were woefully inexperienced college graduates whose sole work experience had been as an intern for a Republican Member of Congress. This was often enough to justify their appointment to tasks such as helping to write the new Iraqi constitution or organizing the privatization of public services. Most US CPA personnel got their first passport to travel to Iraq, and six of the young 'gofers' were 'assigned to manage Iraq's $13 billion budget, even though they had no previous financial management experience' (Chandrasekaran 2007: 104–5).

The liberal peace promotes the individual as the primary unit of society. While such a viewpoint is unproblematic in Western socie-ties, it clashes with many developing world and non-Western contexts in which the family or clan-group may also have significant impor-tance. By empowering individuals (for example, as consumers or as voters with free choice), the liberal peace introduces a cultural clash (sometimes characterized as traditionalism versus modernism) in many societies (see Box 2.1).

If one searches the liberal peace or the operating philosophy of leading states for 'red lines' or non-negotiable elements, then support for the United States seems to be only inviolable principles. Other possible red lines wilt under scrutiny. Commitments to democracy, interna-tional law, human rights or ideas of common humanity ebb according to circumstances. They are upheld in certain cases but conveniently overlooked in others. Even commitments to the inviolability of state sovereignty are abrogated when actors feel strong enough to do so. For example, US drone strikes in Pakistan occur without the support of the Pakistan government. Even the commitment to the free market can be overridden. Despite a rhetoric of austerity and slimming down the state, many cases of Western intervention have involved massive outlays in cash. The US wars of the 2000s have been fought on borrowed money, and compliant regimes – like in Afghanistan and Iraq – have been very costly allies (Bacevich 2010: 246–8). So the free market is not a sacred cow. The leader of Lebanon's Hezbollah,

Box 2.1

A clash of cultures?

One of the authors' postgraduate classes was having a discussion on democratization in post-war societies. The author asked a female Afghan student if she was looking forward to voting for the first time when she returned to Afghanistan. She said that she was not. At this point, the backs of the Western female students visibly straightened. They had been brought up in societies where votes for women was a right and a norm. The idea of an Afghan woman not wanting to vote, when that right had been denied to them for so long, was shocking. But the Afghan student continued, 'My father will decide who I should vote for. So this is not a vote for me. It's an extra vote for him.' She said that she wanted access to decent health care, education and other basic social services, more than participating in an election. Moreover, she did not see a direct link between participating in an electoral process and the delivery of such services. 'Afghan society is clientelistic,' she explained. 'It's all about who you know and not how you vote. For me, voting isn't a "gift" from the West. At best it's an inconvenience.' The Western students were mollified by this explanation but the incident illustrated a clash of cultures. The staging of elections in post-Taliban Afghanistan was more a priority for Western governments anxious to legitimize the rule of their appointed leader than a priority for many Afghans.

Hassan Nasrallah (2008), made an extremely perceptive observation about the red line being compliance with US political objectives:

> The US administration is not concerned with ideologies, beliefs, religions, races and nationalities. Frankly, I tell you, the US administration has no objections should an Islamic party or movement rule any Arab or Islamic country … What counts is your political program … The Americans and Israelis are not concerned with our prayers and fasting. Pray, fast and make pilgrimage to Mecca as much as you want, but leave control and great political interests to be schemed by America and Israel.
>
> (Nasrallah 2008)

The bottom line for Nasrallah was that the US was engaged in simple power politics: protecting its interests and those of its allies. Talk of ideology and principles was erroneous.

If Nasrallah is correct, and the liberal peace has few real principles (despite the rhetoric), then it is worth asking what is the point of the international political and economic system operated by leading Western states. It also explains how radical movements, such as anti-globalization campaigners or even violent fundamentalists in the form of Al-Qaeda or Islamic State, are able to depict 'the West' as an ethical vacuum. Indeed, a number of the martyrdom videos of actual and failed suicide bombers in the United Kingdom mention the baselessness of Western popular culture. Perhaps there is merit to Huntington's (1993) much criticized 'clash of civilizations' thesis, though according to violent fundamentalists, the clash is between an empty mammon-obsessed West and the more sophisticated entities of non-Western identity, religion and customs.

It is very difficult to identify the unifying core behind many Western states. Other than the fact that large numbers of people live there, a rallying theme seems absent. By way of illustration, it is worth looking at speeches by British Prime Minister David Cameron in opposition to radicalization and Islamic extremism. These speeches have much to say about what Britishness is not, but very little about what it actually is (Cameron 2011). The key point is that the combination of economic globalization and a hollow form of liberalism risks producing soulless societies with few centripedal forces. Those with particularist agendas, for example promoting exclusive ethnic or religious views, have the advantage of telling a straightforward story that can capture the imagination and inspire (no matter how illusory it may actually be).

As noted earlier, the manifestations of the liberal peace are most visible in situations of internationally supported peacebuilding and post-war reconstruction. Here the levers of the liberal peace include: Western encouragement (or coercion) to reach a peace deal, direction in writing a new constitution, assistance in establishing and advising political parties, a donors' conference to fund and direct post-war reconstruction, help in holding electoral contests (and sometimes outright interference in the result), programmes of capacity building for the state and civil society, the introduction of 'good governance' targets and the attaching of economic reform conditions to any reconstruction assistance. In effect, the liberal peace has manifested itself in post-peace-accord societies as a conveyor belt of Western inputs. This has resulted in criticisms that it has been formulaic, using a template style of intervention that is unresponsive to the variations demanded by local circumstances (Mac Ginty 2006: 176) (see Box 2.2). According to critics, it is as though IKEA made peace: flat packed and with standardized parts.

The liberal peace is not only restricted to international interventions in societies emerging from civil war. International interventions can

Box 2.2

The viceroy of Bosnia-Herzegovina

Following the 1992–5 civil war, the new state of Bosnia-Herzegovina was ruled first by a NATO and then a European Union interim administration. Certainly, the international community provided the stability and security that was required to pave the way for humanitarian and reconstruction interventions. But critics pointed out that the extent of international intervention by leading states diminished the ability of Bosnia-Herzegovina to stand on its own feet and determine its own course. This view argued that in their desire to protect the peace and the rights of minorities, the international community resorted to paternalistic, even draconian, methods.

Knaus and Martin (2003: 62) summarized the intrusion into Bosnia's affairs thus: 'expatriates make major decisions ... key appointments must receive foreign approval, and ... key reforms are enacted at the decree of international organisations'. The Office of the High Representative, not an elected official from Bosnia-Herzegovina, remains the point of ultimate political authority and has not been afraid to exercise authority in such a way that it has been likened to a 'European Raj'. Until 2005, officials and representatives dismissed by the Office of the High Representative had no legal right of appeal (Zaum 2006: 471). Indeed Chandler (2007: 605) noted that 'sovereignty has in effect been transferred to Brussels.'

In 2008 the International Crisis Group summed up the Western point of view thus: 'Bosnia remains unready for unguided ownership of its own future – ethnic nationalism remains too strong' (BBC 2008). By 2014, however, it had moderated its position. It reflected that 'BiH is trapped in a cycle of poorly thought-out, internationally imposed tasks designed to show leaders' readiness to take responsibility but that put that moment forever out of reach. The only way to encourage leaders to take responsibility is to treat the country normally, without extraneous tests or High Representatives' (ICG 2014: ii).

Sources: Knaus and Martin (2003), Zaum (2006), Chandler (2007), BBC News (2008), ICG (2014).

also be detected in a range of developing world societies. The main means of advance is through economic reform and leverage, and the 'good governance' agenda of bureaucratic reform. Such 'reform' can often sound innocuous, particularly in its promotion of concepts such as accountability, transparency, the regularization of bureaucracy and empowerment. Yet the cumulative effect of such interventions (often multiple interventions by multiple international organizations, bilateral relationships, INGOs and NGOs) may have profound cultural and social impacts capable of influencing the relationships between states, their citizens and markets.

Fundamentally, liberal peace interventions are capable of altering the locus of power within a state. For example, citizens in a clientelist political system may have been used to transferring their allegiance to the ruling party at municipal level in return for resources (Martz 1996). This was a rational transaction for both the citizen and the local political leader, and such relationships would have been deeply embedded into the socio-political culture. Politicians from the ruling party would be routinely invited to family weddings and in return for this public expression of loyalty, the family may expect a share of patronage such as public sector employment for a son (Hamieh 2007). According to the Western mindset, such behaviour may be regarded as 'corrupt' or 'nepotistic' (and indeed, it is often deeply inefficient and patriarchial).

'Good governance' interventions by Western states and international organizations in many developing and post-war contexts have attempted to reform political and economic cultures in order to promote transparency and accountability. In order to build the capacity of municipalities, the United Nations Development Programme may place one of its own personnel in the municipality offices to introduce new administrative procedures. This might involve the ending of discriminatory employment practices whereby loyal supporters received sinecures and protected employment according to who they knew rather than what they knew. The municipality may be unable to refuse the deployment of UNDP staff to its offices because of a directive from central government or because the capacity-building scheme comes with financial inducements. Observers from a Western liberal perspective may applaud the introduction of meritocracy at the municipality. But they may be blind to the impact that such good governance 'reforms' have on altering power relationships within the target society. The citizen may no longer attach legitimacy to the municipality, a factor that is likely to have consequences for political participation, legitimacy and stability. So rather than 'state-building', capacity-building activities funded by the international community might actually help to undermine the state or make apparent state weakness. Chandler (2006: 478) refers to the process as 'the privileging of governance over government'. The key problem is that while people can vote for governments (and government opponents), their votes carry little weight in Washington, Paris, New York or Geneva where important decisions that affect their lives are made.

While many Western liberal interventions are behind the scenes and take the form of bureaucratic meddling, others are much less subtle.

US Vice President Joe Biden visited Beirut ahead of the 2009 Lebanese general election. He made clear that people should vote for the incumbent government or that US aid would cease (Daragahi 2009). Just to make sure that the message was clear, a US naval fleet was parked within sight of the Lebanese coast for much of the election campaign.

The impact of such capacity-building interventions are often quite subtle, and sometimes seem incredibly minor, but their cumulative effect can transform the operating culture and orientation of state institutions (Abrahamsen 2000; Larmour 2005). Whereas a verbal promise between the local mayor and the head of household may have been sufficient in the traditional dispensation, good governance reforms may require the filling in of a standardized form. This apparently innocuous change goes to the heart of some core political and cultural relationships involving issues of trust and reciprocity upon which many societies operate.

It should not be assumed that everything connected with international intervention, the liberal peace and good governance is harmful. Often, for example in the case of state collapse, it is only international actors who are empowered to organize national processes (such as elections) or have the capacity to provide the security necessary for the introduction of humanitarian or development assistance. Since the end of the Cold War, Western-backed interventions have saved and improved lives across war zones and development contexts. Ways of organizing and thinking about international interventions have been inflected with more humane and civilian-orientated sensibilities. The post-Cold War period has seen the development of the human security concept, and innovations in international human rights law (most notably the Responsibility to Protect (R2P) initiative). While these have not delivered on their potential, they have opened space in which the concept of civilian protection has been taken seriously in relation to some conflicts.

Defenders of the liberal peace point out that many underdevelopment and conflict situations would be worse off if there was no international intervention and that we should swallow objections to the imposition of Western values and look to the bigger picture of stability and the potential for economic growth (Quinn and Cox 2007: 518; Paris 2010). Moreover, we should not conceptualize the liberal peace as a dastardly plot solely perpetrated by the United States. Rather than being coerced into the liberal peace, many states, organizations and enterprises see it as a way to further their own ends. France, for example, protested loudly at the prospect of the 2003 US invasion of Iraq. It did not, however, sever its ties with the United States, impose sanctions or take any real steps to prevent the invasion. In short, France had much to gain from its generally good relations with the United States and its connections with the global economy. Other states, even apparent competitors of the US like China, have an interest in stability. As a result, they are often complicit or silent on many foreign policy issues.

We should be under no illusions that international assistance is neutral. It reflects the worldview of those who fund and direct it. Although the liberal peace and the good governance agenda have standardized elements that are applied to different societies with minimal regard to local circumstances, it is important to note that Western interventions and reforms are applied with different levels of enthusiasm in different locations (Richmond 2005a: 217–18). Bosnia-Herzegovina received immense statebuilding and peacebuilding assistance, but the Democratic Republic of Congo did not. International powers pursued Iran to reach a deal on the non-proliferation of nuclear weapons, but have avoided similar activism over Israel and North Korea's nuclear weapons. There were US and UK boots on the ground in Afghanistan, Iraq, Sierra Leone and Bosnia, yet no such enthusiasm in the conflicts in Yemen, Syria and Ukraine. The selectivity in interference is often based on the political mood in the US, UK and other potential troop-contributing countries – not on needs on the ground in the conflict-affected country.

We must be careful not to represent developing world and post-conflict states and communities as mute or powerless actors without agency. In fact, there are many cases of communities resisting, modifying and subverting the original intentions of Western policymakers (Richmond and Franks 2007; Franks and Richmond 2008). Whether in Bosnia, Sierra Leone or Timor-Leste, local political leaders, bureaucrats and communities have found ways to exploit the resources and intentions of international organizations and states involved in peacebuilding and development activities. Indeed an often overlooked, but absolutely crucial, part of the story of international intervention concerns how local bureaucrats bend international edicts to suit their own interests and pace.

While this book is interested in contemporary conflict and development, the parameters of the modern liberal peace are historical. We have had numerous 'defining moments' in which political leaders make 'never again' pronouncements and boldly set a 'new' course: Versailles, Yalta, and the declaration of a New World Order at the end of the Cold War are just a few examples. Yet many of the apparently 'new' structures, institutions and modes of operation seem to resemble those of the previous era (Williams 1998: 5–18). Alex Callinicos (2005: 596) finds that the key to explaining the 2003 Anglo-American invasion of Iraq lies not only in the contingencies of the Bush White House but also in long-maintained historical projects: 'The ideal of a global order in which free markets and democratic institutions promoted peace and prosperity was eloquently articulated by Woodrow Wilson during the First World War.' These continuities guide us towards examining international structures that can survive the tectonic shifts in the international economy and polity.

Alternatives to the liberal peace?

As explained above, the liberal peace is a complex system, with many actors complicit in its maintenance. As a result, the spaces for alternatives to the liberal peace often occur on the margins of societies – in spaces beyond the gaze of leading states and international financial institutions. Hezbollah's creation of a parallel state in Lebanon, along with its own social welfare system, can be regarded as an alternative to a Western-orientated system, but it is by no means a rival to the liberal peace. Similarly, attempts by some Latin American states (most notably Venezuela) to step out from beneath the US shadow may attract headlines but they cannot be regarded as sustainable alternatives to the globalized economy.

The most interesting proto alternative remains China. It is an economic powerhouse (larger than the US according to some measures). But its external politics are difficult to read. It has invested heavily in infrastructure in developing world states, in return for access to natural resources. But it is often mute on many issues, including human rights. It simply gets on with the business of trade and behind-the-scenes influence. It is, however, showing signs of assertiveness in its own 'backyard', the South China Sea, with some commentators fearful that a regional arms race could lead to war. Certainly China's leadership might be tempted to play a nationalist card to avert attention from the lack of domestic freedoms should a Chinese Spring ever get underway. In the meantime, China plays its cards close to its chest and does not overtly threaten the liberal peace.

In order to understand 'structural liberalism' we now turn to the 'hardware' or international institutions that define contemporary conflict and development (Deudney and Ikenberry 1999: 180).

Hardware

States

Rather than provide a 'political science 101' conceptualization of the ideal state, this section will content itself with making three points about the state or the principal unit in the international political system (see Box 2.3). In lieu of an elementary exposition of the concept of the state, readers may wish to consult classics on the subject such as Waltz's (2001) *Man, the State and War,* Migdal's (1988) *Strong Societies, Weak States* or Tilly's (1990) *Coercion, Capital and European States.* The first point is an obvious one: there are enormous variations in the resources, power, legitimacy and capabilities of states and thus in their conflict and development status. Some states are 'lucky' in that they are in a stable region with dependable neighbours, though (as we saw in Chapter 1) states 'fortunate' enough to have

---

Box 2.3

Pakistan: strong or weak state?

Ostensibly Pakistan is a strong state. It covers a huge area (over twice the size of Germany), is rich in resources (including natural gas) and is strategically located (it borders China, India, Iran and Afghanistan). It has an enormous army, is a nuclear power and its economy, though underdeveloped, has experienced strong levels of growth in recent years. Yet the state is chronically weak. It does not control all of its territory (with insurgent groups regularly defeating the national army in the Waziristan region). The bureaucracy is unable to fulfil its stated aims; thus a large proportion of the economy is corrupt and unregulated and – unable to conduct a census – the state can only estimate the population. Democratic institutions are weak, with the result that military coups, states of emergency, mass riots, human rights abuses and political assassinations have been commonplace. Long-running boundary disputes with India, and the Kashmir conflict, have been a drain on state coffers and attentions. So is Pakistan a strong or weak state? The answer, despite its formidable nuclear arsenal, must be that it is weak. The state faces too many alternative sources of power to be able to assert itself.

natural resources are often cursed with conflict over the distribution of these resources. The massive variance in state capability leads to uneven international institutions, with some states (and corporate bodies and lobby groups) wielding disproportionate influence. In stark terms, while some states and lobby groups can afford to maintain extensive bureaucracies at international organizations, other states cannot (Robbins 2003: 105). Indeed, Japan is reported to pay the International Whaling Commission membership fees for some states in return for support on the right to hunt whales (*Sunday Times* 2010).

State capacity is a crucial factor in facilitating, developing and constraining conflict. Adrian Leftwich's (1996: 284) concept of the 'developmental state' is useful in illustrating how states need to 'concentrate sufficient power, authority, autonomy, competence and capacity at the centre to shape, pursue and encourage the achievement of explicit developmental objectives'. As Leftwich (1996) points out, states have pursued this capacity to develop in very different ways, with Taiwan and South Korea achieving very high rates of growth under authoritarian political systems. Other states simply did not have the capacity or inclination to engage in national mobilization for economic growth.

This leads us to the second point: that the type of state matters, particularly in terms of its internal political organization. Some forms of state organization may be more conducive than others to promoting development and constraining conflict. Democracy and development may not always be compatible, and it is unlikely that China could achieve such high economic growth rates (an average of 9 per cent 1989–2015) if it were embarking on a serious programme of political liberalization (Trading Economics 2015). The state is often the central clearing house for societal conflict, providing rules and mechanisms to allow individuals and groups to coexist. But it may also be inept or incapable, or may have little interest in maintaining its conflict-regulating responsibilities. William Reno (1997) paints a dystopian picture of West African states in the late 1990s, in which ruling elites regarded citizens as an expensive encumbrance and thus made no pretence at offering public services. Instead, they fixed their energies on the extraction of precious minerals, usually in concert with external commercial interests. In some cases, especially societies with identity fissures, the state may privilege some groups and discriminate against others. An enormous literature posits a link between democracy and peaceful relations between states, but the relationship is by no means simple, and this 'democratic peace' literature tends to

overlook the locus of most violent conflict: within the state itself
(Doyle 1980; Henderson 2002: 2).

A key point to bear in mind is that in many societies the state is the
only political and economic prize worth having. To be excluded from
the state means to be cut off from virtually all public resources.
Indeed, one can see this quite literally in parts of Africa where the
road and electricity pylons stop abruptly because a district did not
support the ruling elite. Ian Taylor (2005) observes that

> control of the state serves the twin purposes of lubricating patronage
> networks *and* satisfies the selfish desire of elites to enrich themselves,
> in many cases in quite spectacular fashion. That is what lies at the heart
> of the profound reluctance by African presidents to hand over power
> voluntarily and why many African regimes end messily, often in coups.
>
> (Taylor 2005: 4)

He goes on to note how neo-patrimonialism or clientelistic 'big man'
politics is alive and well across the region. This is despite enormous
democratization and 'good governance' interventions over many
years. In other words, a political culture may be surprisingly resilient
regardless of externally promoted institutional engineering.

The third point to make in relation to the state is to note the international community's near addiction to statebuilding as the standard
response to civil war or political transition. It seems as though the only
tool in the toolbox is state rebuilding (as manifested through the already
mentioned capacity-building programmes, good governance, public
sector reform, etc.). In development contexts, Kenny (2003) notes:

> The state is either the solution, the only way to combat structural
> weaknesses that hold back growth, or it is the problem, tying down
> the invisible hand; or it is the facilitator, vital for the efficient func-
> tioning of the free market.
>
> (Kenny 2003: 413)

Either way, the state is a key ally or enemy in international efforts to
promote development or reduce conflict.

Failed states

In some parts of the world, the state is patently a dysfunctional political
and economic model. Terms like 'failed' or 'fragile' state are laden

with baggage, and some have even argued that the US is a failed state, given its permanent budget deficit and massive prison population (Whitmeyer 2013). But the US can wield immense power in a controlled way. Other states, like Yemen, Iraq, Libya, Syria, Nigeria or Ukraine, do not control all of their territory. In entire regions, like the Horn of Africa or Central and West Africa, the state has not taken root. In fact, for many people, the state is irrelevant in daily life (Kabamba 2010). People are left to get on with their own survival and prosperity while the state does little in terms of providing public good like health care or security. The state may manifest itself occasionally in the shape of a police car visiting the village every few weeks, but otherwise it might be absent, occasional, irrelevant or incompetent.

Despite abundant evidence that some states just don't work, the international community remains addicted to the state as the primary unit of political organization. Thus immense international effort is invested into statebuilding and state reform. The phrase attributed to the car manufacturer Henry Ford, that a customer can have a car painted any colour – as long as it is black, can be remodelled to read that any political unit is tolerable – as long as it is a state. The fear of statelessness (regarded by many as a deviant form of political organization) is very grave indeed. According to one observer, 'Failed or failing states are often Petri dishes for transnational criminal activity such as money laundering, arms smuggling, drug trafficking, people trafficking, and terrorism' (Wainwright 2003: 486).

While huge international resources are poured into shoring up and reconstructing states (Mayall and Soares de Oliveira 2011), it is worth noting that the international system is selectively tolerant of different types of state. This tolerance often depends on three factors: the ability of a state to resist or subvert Western influence (for example, a powerful China), the strategic importance or unimportance of the state (for example, oil-rich Iraq or strategically marginal Haiti), and the points of economic and geopolitical confluence between the state and leading states in the international political system (for example, Israel's ability to organize a strong lobby in the US). Good examples of patently 'bad' but tolerated states come in the form of the oil-rich Gulf monarchies that can at best be described as authoritarian and controlling, and at worst as despotic and tyrannical. Saudi Arabia executed over 100 people in January to June 2015, most for non-violent offences (Human Rights Watch 2015a). Its campaign to deport undocumented workers has led to mass human rights abuses (Human

Rights Watch 2015b). Its airstrikes on the Houthi group in Yemen involved the rise of cluster munitions and resulted in hundreds of civilian casualties (Human Rights Watch 2015c). Women cannot vote, drive a car or leave the house without a chaperone (*The Week* 2015). By most measures, Saudi Arabia is a state with scant regard for basic human rights. Yet, when long-term Saudi monarch King Abdullah died in early 2015, UK Prime Minister David Cameron ordered that flags be flown at half mast on government buildings throughout the UK (Sparrow 2015). A predecessor Prime Minister, Tony Blair, intervened to stop a fraud investigation into BAE Systems and its arms deals with the Saudis (Leigh and Evans 2010). John Sawers (2015), the former head of the UK's spy agency and thereafter a board member of British Petroleum, wrote in the *Financial Times* how Saudi Arabia was the best chance for order in the region and was on the road to meritocracy. His evidence was nugatory.

It is clear that the geo-strategic concerns of leading Western states outweigh the desire to upset regional power-holders (Ehteshami and Wright 2007). The worldviews of the ruling elites in the Gulf region are congruent with Western political and economic interests and so their unsavoury political backyard is spared scrutiny and intervention. Gulf states invest their oil wealth in Western economic markets, buy prodigious quantities of Western arms and are on the 'right side' in the War on Terror and the Sunni versus Shiite struggle. In one of the many contradictions that defines the international political system, this Western laissez-faire attitude to capable and compliant Gulf states contrasts with 'failed' or 'failing' weak states that are not in a position to resist Western intervention. As George Orwell might have put it: some states are more equal than others.

International organizations

That international organizations exist at all is remarkable. The jealousies of national interests, as well as the fallout from international crises, have meant that the international institutions of previous eras often resembled short-lived tactical alliances rather than permanent forums for the regulation of international society. Even more remarkable has been the existence of some international institutions for so many decades (the United Nations, World Bank and International Monetary Fund for over seven decades, and the European Union – albeit in a greatly modified form – for over six decades).

The United Nations

As the planet's premier collective security organization, the United Nations is the target of immense criticism. Common chants are that it's too inefficient, bureaucratic, slow, corrupt, under funded and unwieldy (MacFarlane and Khong 2006; Urquhart and Weiss 2012). Yet, the organization is merely the sum of its parts (member states) and is a largely accurate reflection of states' relationships with one another and their attitudes towards pooling sovereignty for the collective good. That it tends to be reactive rather than proactive, is highly selective in its interventions, and is inconsistent in its attitude towards state sovereignty is not the fault of international mandarins working in a vacuum in New York and Geneva. Instead, it is the fault of the national governments who have created and maintained the system.

While it is easy to criticize the United Nations, it is also easy to overlook the immense (and often unsung) development, conflict-amelioration and humanitarian work it undertakes. Often this work is conducted through specialist agencies, has long-term impacts, and makes a qualitative difference to the lives of millions in underdeveloped and war-torn states. Thus UN agencies are responsible for the physical security, legal protection, nourishment, shelter and repatriation of substantial numbers of individuals and communities: a fact that is routinely overlooked by those who use broad-brush criticisms against the organization.

The United Nations has developed as a result of disjointed incrementalism. Rather than planned strategic growth, UN capabilities have developed reactively in the face of crises, with the end of the Cold War simultaneously lifting an immense constraint on its ability to operate and presenting it with a vastly increased workload. It responded in the 1990s by having ever more ambitious operations in an increased number of theatres (Bellamy and Williams 2010: 93–120). The growth in peacekeeping operations occurred despite the UN Charter not mentioning the term 'peacekeeping'. Nor does the UN have its own army; it relies on member states to contribute troops. In the post-Cold War period there has been a trend towards 'outsourcing' or 'subcontracting', with the employment of regional organizations, INGOs and NGOs to carry out peace-support and development activities on behalf of the United Nations (Richmond and Carey 2005). A privatization of UN functions has also been visible, with – among other trends – a growth in the use of private security companies to provide security for UN personnel (Pingeot 2012: 7)

On the positive side of the ledger, the UN's post-1990 operations showed that the organization was adopting a more sophisticated understanding of conflict, especially with regard to the complexities of the relationship between conflict and development. As of mid-2015, the organization has 106,000 personnel serving in 16 peace-keeping missions – most of which are in Africa (Williams 2011: 185–92). On the negative side of the ledger, there was discomfort at the more robust aspects of some peacekeeping interventions in which peace 'keeping' became peace 'making' and peace 'enforcing' (Boulden 2001). Such qualms echoed those over the notion of 'humanitarian war' and concerns that humanitarian interventions could prolong and intensify war (Janzekovic 2006; Belloni 2007; Bellamy and Williams 2010: 229). There were also accusations of 'mission creep', whereby originally modest UN interventions became victim to ever-broadening mandates. This was especially the case in missions that required extensive nation and statebuilding, activities that are necessarily long term and expensive (Pugh 1997). The term 'mission creep' came to prominence in relation to an originally modest UN effort in Somalia in the early 1990s. What was originally an attempt to secure supply routes for humanitarian aid convoys led to direct conflict between UN contingents and Somali warlords.

The United Nations has undergone significant reform, and embarked on major initiatives, since the 1990s. Its eight Millennium Development Goals helped inject focus into development interventions undertaken by it and its member states, and are due to be replaced by

Plate 6 *A UN 4×4 vehicle: the UN is the planet's premier collective security organization.*

17 Sustainable Development Goals, with 169 targets to be met by 2030. The 2001 Brahimi Report, although not fully implemented, recognized many of the UN's organizational shortcomings. The 2003–6 High-Level Panel on Threats, Challenge and Change considered the new range of transnational threats faced by states and the utility of UN structures and practices to deal with them (Hannay 2005; Stedman 2007). The 2005 establishment of a Peace Building Commission signalled a willingness among member states to take on board the lessons from its previous and ongoing peace-support operations. A 2015 review sought to 'propose ways to strengthen the performance and impact of the Peacebuilding Architecture, with a view to realizing its full potential' (United Nations 2014).

The principal problem facing the United Nations is the perennial struggle between state sovereignty and the collective good. On many occasions the two are simply incompatible. Neo-conservative elements in the United States are at least honest when they voice their suspicions about multilateralism or the potential of any national or multilateral power source to rival its hegemonic position (Callinicos 2005: 598). As the 2003 Anglo-American invasion of Iraq and its bloody aftermath showed, leading states are willing to override the UN when it suits, but not above appealing for UN assistance when they find themselves in a mess. Important decisions on restructuring the Security Council to make it more representative have been dodged and it is likely that the pattern of disjointed incrementalism will continue. The permanent five members of the UN Security Council – US, China, Russia, UK and France – reflect the world in 1945. There are few credible reasons for excluding powerful populous states like Brazil or India from the Security Council other than the concentration of power in the hands of those that already have it.

Ultimately the UN relies on its moral authority – a sense that a collective security organization represents world opinion. This moral authority, to the extent that it ever existed, is under severe pressure. States are wary of the power they are willing to cede to the UN. They are content for relatively anonymous figures to be appointed Secretary General. A good indication of the importance placed on the UN in the policing of world events is the number of troops and police states are willing to provide for UN peacekeeping missions. In May 2015, the Russian Federation, with an armed force of over 1.2 million, had 71 personnel serving on UN missions. The figure for the US (with an armed force of over 1.4 million) was 80 (World Bank 2015).

Peacekeeping has become a developing world activity, with countries like Cameroon (1380 personnel serving with the UN), Senegal (3575 personnel) and Indonesia (2729 personnel) taking the strain of providing troops for UN missions (United Nations Peacekeeping 2015).

Other international organizations

Regional security and economic organizations have shown themselves to be increasingly capable of pursuing conflict-amelioration and development agendas. The African Union (AU) is comprised of all African states, except Morocco. It was established to replace the ineffective Organization for African Unity, and was partly a response to the 1994 Rwandan Genocide and the need for an African response to sudden-onset emergencies (Plant 2014). It has steadily expanded its operations to include peacekeeping, election observation and humanitarian assessments but it is hobbled by capacity and political-will issues (Renwick 2015). It is not clear that the AU promise of adopting 'African solutions for African problems' has been met. Its definitions of peacebuilding, for example, seem to be cut and pasted from Western sources and overlook indigenous and traditional forms of dispute resolution that might be found closer to home (Mac Ginty 2008). An indicator of continent-wide commitment to the AU comes from the fact that its new headquarters in Addis Ababa was a gift from China, and only about 40 per cent of the organization's costs are met by African states. The rest is made up of donations by the UN, EU and major donors (Al Jazeera 2013).

NATO and the European Union are probably examples of the most capable regional security organizations. In part, this is because of the relative wealth of their members, and their alignment with major states with a record of intervention. The EU has beefed up its foreign policy capabilities with its External Action Service to give weight to the commitments of a Common Foreign and Security Policy. The EU has engaged in approximately 30 overseas military and civilian operations, usually in cooperation with NATO, the UN or AU. One mission worth mentioning is the EU Monitoring Mission in Georgia. A response to the 2008 Russia–Georgia conflict, it is small, unarmed, mainly civilian and was placed in the field very quickly. Like other international organizations, member states are wary of giving up power to the EU, but it is a major player in peace, development and emergency response.

After looking as though it had no role in the post-Cold War world, NATO's future as a security provider is secure. In part this is because of concerns over a more confident Russian Federation. But it is also because NATO has redefined itself as an organization able to operate 'out of area' – that is, outside of the European theatre of operations it was designed for. NATO has deployed to Afghanistan, Iraq, Somalia and counter-insurgency operations off the Horn of Africa.

An interesting development has been the 2010 establishment of the G7+ group of fragile and conflict-affected states. Comprised of 20 states (among them South Sudan, Haiti and Yemen), it was a reaction to the top–down nature of much development and peacebuilding interventions. The states involved realized that they could benefit from sharing experiences among themselves and lobbying as a group. Their main initiative thus far has been agreeing on the New Deal for Engagement with Fragile States (OECD 2011), a manifesto that seeks to shape peacebuilding and statebuilding activities from within conflict-affected countries. Members committed to 'focus on new ways of engaging, to support inclusive country-led and country-owned transitions out of fragility'. It is too early to tell if the G7+ and the New Deal constitute what the World Bank optimistically declared as a 'paradigm shift' in how peace and development were managed (World Bank 2011; Woollard 2013). As it is an intergovernmental organization, its members face the temptations of endless summit meetings that reproduce rather than challenge existing international order.

International non-governmental organizations

INGOs have transformed international responses to conflict and development since the 1970s. Organizations such as the International Committee of the Red Cross and Red Crescent, Doctors Without Borders or Oxfam have been important for at least three reasons. First, they have allowed official development assistance, conflict prevention and peacebuilding activities to have a much further reach than traditional bilateral or international organization activity. INGOs are often cheaper and more flexible than official modes of aid delivery, particularly in cases where they subcontract to local NGOs (Richmond and Carey 2005). Second, INGOs have prompted more cases of development intervention by leading states and international organizations. This has been particularly the case in their ability to 'bear witness' and engage in advocacy, thus bringing humanitarian emergencies and chronic development and conflict situations to the attention of publics and polities. The third, and most significant, point has been the ability

of INGOs to play a role in shaping the development, conflict prevention and development assistance agenda of leading states, leading
international organizations and international financial institutions.
This influence has been by no means total, nor has the process been
one way, yet INGOs have been incredibly important in broadening
governmental understanding of conflict and making obvious the multiple connections between conflict and development.

Some INGOs have been co-opted into the liberal peace project.
Rather than acting as bulwarks against the ambitions of rich Western
states, some INGOs are the principal transmission agents of the liberal peace through their promotion of good governance and reform
agendas (Richmond 2005b). The privatization of development and
humanitarianism (for example, through tender processes for development contracts) leaves many INGOs with little choice. To stay in
business (and to stay relevant and to fund their other advocacy work)
they must compete for contracts released by governments and international organizations. Yet all contracts come with conditions, and
some of these conditions may constrain the original development and
conflict-amelioration objectives of the INGO. As Box 2.4 shows,
increasing privatization in the humanitarian and development sectors
has raised many practical and ethical questions.

Box 2.4

Privatizing security, humanitarianism and development

The private sector is often considered to be more efficient and cost-effective than
state-run organizations. The rigour of the private sector has been increasingly applied
to the provision of security, humanitarian and development interventions in some contexts as donor states and organizations seek to limit costs and devolve responsibilities.
Private military contractors (PMCs, formerly called 'mercenaries') have been
employed in Iraq and Afghanistan to protect embassies and reconstruction projects.
Controversy has abounded, especially in relation to the immunity from prosecution
enjoyed by some PMCs in some contexts (Baer 2007). For states sensitive to headlines
of body-bags coming home, non-state partners (whether charitable or private) are an
attractive proposition. A number of scholars and aid practitioners, however, have
pointed out that the apparent privatization and securitization of humanitarianism and
development are leading to foundational changes in how we conceptualize 'charity',
'aid', 'neutrality' and 'humanitarianism' (Duffield 2007; Spearin 2008).

Sources: Baer (2007), Duffield (2007), Spearin (2008).

International financial institutions

UN and INGO personnel are often highly visible in conflict and development environments through their blue helmets, white 4×4s and prominent banners advertising their projects. Yet one set of highly influential international actors is frequently less visible: the international financial institutions (IFIs). The World Bank (originally the International Bank for Reconstruction and Development) and International Monetary Fund date from the era when Adolf Hitler was still in power. The framers of both organizations were conditioned by their experiences of the 1930s, with the contraction of world trade and state attempts to 'beat' economic depression through tariffs. Although the IMF and World Bank are often discussed in the same breath, and although both share the same aim of facilitating free trade, they are 'not identical twins' (Hanlon 1996: 25). The IMF's chief role is to smooth international trade through short-term balance of payments assistance to states. Its main focus is on macro-economic matters and it regards low inflation (achieved by tight government spending regimes) as the most important weapon in the regularization of trade and the stabilization of currencies (Pollard 1997: 79). The World Bank has a broader brief, and is focused on development and longer-term interventions. It aims to encourage states to change the structure of their economies so that growth and trade can operate unhindered. A welter of associated bodies and permanent conferences (the General Agreement on Tariffs and Trade, the World Trade Organization, G8, the United Nations Conference on Trade and Development and the Uruguay and Doha Rounds of negotiations, etc.) assist in the overall aim of promoting free trade.

As was made clear in the Introduction to this book, the international financial architecture has been heavily criticized as being designed to perpetuate the poverty and the disadvantaged positions of developing world states (Stiglitz and Charlton 2005). Thus, the IMF and World Bank are routinely demonized as being 'the high priests of neo-classical orthodoxy' (Shutt 1998: 160) or progenitors of 'predatory globalisation' (Falk 1999). The reasons for this derision are clear: the international financial architecture has facilitated the post-Second World War growth in the disparity between rich and poor states and the income disparities in developing states (Seligson 2003: 2). Falk (1999: 13) characterizes the international economy as structurally racist: it is an 'economic apartheid' that reinforces a rich white world and is comfortable with the impoverishment of the non-white world. Joseph Stiglitz (2003), for many years an economist with the IMF, pursues a similar theme: the IFIs prioritize the interests of global capital above those of global community.

Given our interest in the international financial architecture is in how it facilitates and thwarts conflict and development, three points can be made. The first point is that the IFIs are here to stay. Although many commentators paint them as pantomime villains (with considerable justification), there are no alternative sources of international economic regulation waiting in the wings. The IFIs have undergone considerable change throughout their history and have presided over many economic downturns (for example, the 1973 oil and dollar shocks, the 1994–5 Mexican peso crisis and the 1997–8 Asian financial crisis, the 2007–9 market instability and credit crunch). Yet, for all of this mismanagement and the disjointed incrementalism of their institutional evolution, there is no political appetite among leading states for root and branch reform (Harris 1999; Underhill 2001: 291). Despite the iniquities of the situation, societies emerging from civil war or suffering from underdevelopment simply have to work with the IFIs. States emerging from civil wars are unable to avail themselves of reconstruction loans or to conduct international trade unless they sign up to IFI strictures. These strictures may include honouring the external debts of the previous regime, even if that regime lacked a democratic mandate.

There is a new rival in the form of the Asian Infrastructure Investment Bank (AIIB), a Chinese-backed initiative that also has the support of the other BRICS (Brazil, Russia, India, China and South Africa). It is too early to say whether it will succeed, but it is likely to become a way of allowing China to expand its soft power by lending developing countries sums to build infrastructure (Perlez 2014).

The second point concerns the worrying absence of democracy and transparency in the affairs of the IFIs. There is a certain irony here in that while democracy and democratization are often pushed on states emerging from conflict as the route to guarantee them stability, these states must also enter into deeply undemocratic relationships with the IFIs. Joseph Hanlon (1996: 24) noted how, in the aftermath of a civil war, Bretton Woods officials in Washington had 'more power in Mozambique than any Mozambican, up to and including President Chissano'. Unelected IFI officials have been present at many civil war peace negotiations: they are difficult guests to refuse in that reconstruction assistance is often routed through them. While the World Bank, IMF and other international agents encourage transparency among recipient states, the World Bank's own lack of transparency was illustrated by the controversy surrounding a massive pay rise awarded to its president's girlfriend in 2005. Decision-making power in the IMF is based purely on financial clout, with states enjoying voting rights commensurate with their contributions to the Fund. As a result, the poorest states have little say in how the Bank operates.

The third point is that IFI strictures can have profound impacts on social harmony in the aftermath of a civil war and the ability of a post-civil-war government to fulfil its social provision pledges. This can have serious consequences, since the post-civil-war political dispensation will require a broad base of stakeholders if the peace accord is to hold. IMF and World Bank mantras of 'small government' can severely restrict the ability of governments to honour the 'peace dividend' pledges that accompanied a peace accord. As Addison (2003: 264) notes, 'Decisions made in the early phase of the war-to-peace transition alter the distribution of productive assets and thereby the benefits of growth to the rich and the poor for years to come.' The 2006 Nepalese peace agreement, for example, contains a highly ambitious list of social pledges that would make a 1970s Scandinavian socialist government proud. Yet the Nepalese state does not have autonomy over how it disburses the contents of the public coffers. While the IFIs do not have peacekeepers or storm troopers on the battlefields, they are not entirely removed from the dynamics that encourage conflict and development.

Conclusions

To return to the software and hardware metaphor, it would seem that an incredibly sophisticated software prevails in the form of the liberal peace. Through a mixture of subtle and not-so-subtle inducements and compliance mechanisms, states are encouraged to conform to an order that reinforces existing power-holders. Its chief aims are to stabilize dysfunctional war-torn states, to promote liberal economics, and to perpetuate the control of leading states (Duffield 2001: 10). The liberal peace regime allows some wriggle room: it has been softened by the human security perspective, will face increasing challenges from a more internationally active China and – in certain circumstances – can be subverted by states and communities. While the dominant software is constantly upgraded, the hardware (in the form of international institutions) is clunky, slow and often fails to work. In a sense, the software has found ways in which it can work without hardware, or certainly without the original hardware.

The institutional factors that shape contemporary conflict and development are undergoing simultaneous processes of continuity and change. The forces of continuity are to be found in the rigidity of a UN Security Council that reflects the 1945 balance of power and in the inflexibility of state sovereignty as interpreted by some states. But there are many signs of change, all of which impact on the ebb and flow of conflict and development and the ability of international institutions to react. The range of factors deemed capable of causing conflict (or instability) and thwarting economic and human development is deemed to have grown in the post-Cold War era. The transnational nature of many of these threats (for example, legal and illegal labour flows or environmental degradation) means that traditional state-centric institutions are often poorly equipped to deal with them. While many international institutions have been changing to anticipate or react to a new range of threats, it is not at all clear that the current suite of collective security organizations and development organizations is the best equipped to deal with these threats. It is an incontrovertible fact that none of the current international or regional security organizations in existence was explicitly formed to deal with contemporary civil war or underdevelopment. Quite simply, the international political system is stuck with the wrong tools for the job.

Having said that, many international organizations have undergone processes of reform and there are growing interconnections between the traditional international units (such as states and collective security organizations) and more modern units in the form of INGOs and organized global capital. This 'complex multilateralism' allows numerous multilateral associations to come together, often on specific issues and for limited time periods. Thus campaigns to address blood diamonds, child soldiers or landmines will see 'mixed-actor coalitions' of states, international organizations, pressure groups and private sector bodies cooperate on a single issue.

There seems to be no end in sight to the perennial problem of reconciling the national interests of states and the need for states to put aside narrow national interests to deal with pressing collective problems such as global warming or poverty. International organizations are the clearing house for many of these tensions. In some cases, such as agreement on the Millennium Development Goals, states have been able to cooperate on joint aspirations. Logjams in international organizations, however, have tempted some states and interests to act either unilaterally, or multilaterally, though not through an international organization. Some commentators have suggested the need for a 'league of democracies' or an alliance of Western democratic states who could act without the encumbrance of an unwieldy international organization (Kagan 2008). The advantage would be swiftness of action and unity of purpose in the face of a humanitarian or complex political emergency. Critics, however, have pointed out that such a league would amount to a charter for interventionism and would conveniently free Western states of the need to persuade non-Western states of the merits of their case.

Summary

- Rather than simply look at international institutions, it is also important to look at the principles and values behind the institutions.
- The concept of 'the liberal peace' has been used to describe the peace- and development-support interventions by leading states, leading international organizations and international financial institutions in war-affected and developing societies. According to critics, the liberal peace is guilty of Western bias and can be a form of neocolonialism.
- While good governance reforms might seem wholly sensible to Western eyes, they can have a profound effect on important relationships in target societies. For example, they might alter how citizens interact with and perceive the state and fellow citizens.
- The international financial institutions wield immense economic and political power, but it is difficult to think of alternatives to them.
- The international political system increasingly operates according to 'complex multilateralism' whereby a constantly shifting mix of states, international organizations, INGOs, NGOs and private sector interests coalesce and recoalesce at different times on different issues.

Discussion questions

- If the United Nations collapsed, what would you replace it with? What principles should underpin any new collective security organization?
- Is there any way around the tension between the national interests of states and the collective needs of a community of states?
- Are there any serious alternatives to the state as the basic political unit in the international system?
- What advantages can private sector firms bring to security, humanitarian and development interventions?
- Has Responsibility to Protect (R2P) worked?
- In 2015, during a visit to Kenya, US President Obama chided the Kenyan government for its laws that ban homosexuality. Kenyan President Kenyatta replied: 'There are some things that we must admit we don't share. It's very difficult for us to impose on people that which they themselves do not accept.' Was President Obama correct to impose one culture on another?

Further reading

M. Douglas (1986) *How Institutions Think*, Syracuse, NY: Syracuse University Press provides a seminal account on the inner workings of large organizations. Insider's guides to the role of the IMF and US capital in the developing world can be found in J. Stiglitz (2003) *Globalization and its Discontents*, New York: Norton, and J. Perkins (2004) *Confessions of an Economic Hit Man*, San Francisco, CA: Berret-Koehler. An extremely good survey of some of the problems facing INGOs and NGOs is H. Yanacopulos and J. Hanlon (eds) (2005) *Civil War, Civil Peace*, Oxford: James Currey. Revealing insights into the nature of Western interventions can be found in S. Autessere (2014) *Peaceland: Conflict resolution and the everyday politics of international intervention*, Cambridge: Cambridge University Press, R. Chandrasekaran (2007) *Imperial Life in the Emerald City: Inside Iraq's Green Zone*, London: Bloomsbury, and T. Rick (2006) *Fiasco: The American military adventure in Iraq*, London: Penguin. A wonderfully honest (and funny) reminder of how Western preconceptions sometimes travel poorly can be found in N. Barley (1983) *The Innocent Anthropologist: Notes from a mud hut*, London: British Museum Publications.

Useful websites

Most international organizations, international financial institutions and INGOs have content-rich websites. Major international organizations include the United Nations (www.un.org), the European Union (http://europa.eu/), the African Union (www.Africa-union.org), the Organization of American States (www.oas.org), the League of Arab States (www.arableagueonline.org) and the South Asian Association for Regional Cooperation (www.saarc-sec.org). Some of these organizations have agencies or sections with a specific remit for conflict and/or development. For example, the European Commission's Humanitarian Aid Office (http://ec.europa.eu/echo) and the UN's Peacebuilding Commission (www.un.org/peace/peacebuilding) or Department of Peacekeeping Operations (www.un.org/Depts/dpko). The World Bank (www.worldbank.org), IMF (www.imf.org), Asian Development Bank (www.adb.org) and European Bank for Reconstruction and Development (www.ebrd.com) are all represented online. There is much data available at the individual state level. See, for example, the Fragile States Index (http://fsi.fundforpeace.org/), or the Global Peace Index (visionofhumanity.org) and reports by Freedom House (https://freedomhouse.org/) and Amnesty International (www.amnesty.org). Brettonwoodsproject.org keeps an eye on the IMF and World Bank. Many INGOs have very useful resources in terms of reports and primary documents. See, for example, sipri.org; international-alert.org; accord.org.za; and c-r.org.

References

Abrahamsen, R. (2000) *Disciplining Democracy: Development discourse and good governance in Africa*. London: Zed.

Abrahamsen, R. (2004) Poverty reduction or adjustment by another name? *Review of African Political Economy* 31(99): 184–87.

Addison, T. (2003) Communities, Private Sectors, and States. In Addison, T. (ed.) *From Conflict to Recovery in Africa*. Oxford: Oxford University Press, pp. 263–87.

Al Jazeera (2013) African Union seeks financial independence. Al Jazeera. 24 May. Available at: http://www.aljazeera.com/news/africa/2013/05/201352412928567270.html (accessed on 8 September 2015).

Bacevich, A.J. (2010) *Washington Rules: America's path to permanent war*. New York: Metropolitan Books.

Baer, D. (2007) The immorality of Blackwater. *Guardian*, 6 October.

BBC News (2008) Country Profile: Bosnia-Herzegovina. *BBC News Online*. 1 January. Available at: http://www.bbc.co.uk/news/world-europe-17211415 (accessed on 7 March 2008).

Bellamy, A. and Williams, P. (2010) *Understanding Peacekeeping*, 2nd edn. Cambridge: Polity.

Belloni, R. (2007) The trouble with humanitarianism. *Review of International Studies* 33: 451–71.

Boulden, J. (2001) *Peace Enforcement: The United Nations experience in Congo, Somalia, and Bosnia*. Westport, CT: Praeger.

Callinicos, A. (2005) Iraq: fulcrum of world politics. *Third World Quarterly* 26(4–5): 593–608.

Cameron, D. (2011) Speech on radicalization and Islamic extremism, *New Statesman*. 5 February. Available at: http://www.newstatesman.com/blogs/ the-staggers/2011/02/terrorism-islam-ideology (accessed on 8 September 2015).

Campbell, S., Chandler, D., Sabaratnam, M. (eds) (2011) *A Liberal Peace? The problems and practices of peacebuilding*. London: Zed.

Chandler, D. (2006) Back to the future? The limits of neo-Wilsonian ideals of exporting democracy. *Review of International Studies* 32(3): 475–94.

Chandler, D. (2007) EU Statebuilding: securing the liberal peace through EU enlargement. *Global Society* 21(4): 593–607.

Chandrasekaran, R. (2007) *Imperial Life in the Emerald City: Inside Iraq's Green Zone*. London: Bloomsbury.

Craig, D. and Porter, P. (2003) Poverty reduction strategy papers: a new convergence. *World Development* 31(1): 53–69.

Daragahi, B. (2009) Joe Biden, in Lebanon, hints that US might cut aid if Hezbollah wins election. *Los Angeles Times*, 23 May. Available at: http://articles.latimes. com/2009/may/23/world/fg-biden-lebanon23 (accessed on 8 September 2015).

Deudney, D. and Ikenberry, G.J. (1999) The nature and sources of liberal international order. *Review of International Studies* 25(2): 179–96.

Doyle, M. (1980) Liberalism and world politics. *American Political Science Review* 80(4): 1151–69.

Duffield, M. (2001) *Global Governance and the New Wars: The merging of development and security*. London: Zed.

Duffield, M. (2007) *Development, Security and Unending War: Governing the world of peoples*. Cambridge: Polity.

Ehteshami A. and Wright, S. (2007) Political change in the Arab oil monarchies: from liberalization to enfranchisement. *International Affairs* 83(5): 913–32.

Falk, R. (1999) *Predatory Globalization: A critique*. Cambridge: Cambridge University Press.

Franks, J. and Richmond, O. (2008) Coopting liberal peace-building: untying the Gordian knot in Kosovo. *Cooperation and Conflict* 43: 81–103.

Hamieh, C. (2007) Intra-party competition in a divided society: Amal versus Hezbollah in Lebanon, 1998–2003. Unpublished doctoral thesis, University of York.

Hanlon, J. (1996) *Peace Without Profit: How the IMF blocks rebuilding in Mozambique*. Dublin: Irish Mozambique Solidarity and the International African Institute in association with James Currey and Heinemann.

Hannay, D. (2005) Reforming the United Nations: The use of force to safeguard international security and human rights. A member's perspective from the Secretary-General's High Level Panel on Threats, Challenges and Change. *Conflict, Security and Development* 5(1): 109–17.

Harris, L. (1999) Will the Real IMF Please Stand Up? In Michie, J. and Grieve Smith, J. (eds) *Global Instability: The political economy of world economic governance*. London: Routledge, pp. 198–211.

Henderson, E.A. (2002) *Democracy and War: The end of an illusion?* Boulder, CO: Lynne Rienner.

Human Rights Watch (2015a) Saudi Arabia: 100 executions since 1 January. Human Rights Watch, 16 June. Available at: https://www.hrw.org/news/2015/06/16/saudi-arabia-100-executions-january-1 (accessed on 8 September 2015).

Human Rights Watch (2015b) Saudi Arabia: Mass expulsions of migrant workers. Human Rights Watch, 19 May. Available at: https://www.hrw.org/news/2015/05/09/saudi-arabia-mass-expulsions-migrant-workers (accessed on 7 August 2015).

Human Rights Watch (2015c) Yemen: Unlawful airstrikes kill dozens of civilians. Human Rights Watch, 30 June. Available at: https://www.hrw.org/news/2015/06/30/yemen-unlawful-airstrikes-kill-dozens-civilians (accessed on 7 August 2015).

Huntington, S. (1993) The clash of civilizations? *Foreign Affairs* 72(3): 22–49.

International Crisis Group (2014) Bosnia's Future, Europe Report 232, 10 July. Available at: http://www.crisisgroup.org/~/media/Files/europe/balkans/bosnia-herzegovina/232-bosnia-s-future.pdf (accessed on 8 September 2015).

Jacoby, T. (2007) Hegemony, modernisation and post-war reconstruction. *Global Society: Journal of Interdisciplinary International Relations* 21(4): 521–37.

Janzekovic, J. (2006) *The Use of Force in Humanitarian Intervention: Morality and practicalities*. Aldershot: Ashgate.

Kabamba, P. (2010) Heart of Darkness: current images of the DRC and their theoretical underpinnings. *Anthropological Theory* 10(3): 265–301.

Kagan, R. (2008) The case for a league of democracies. *Financial Times*, 14 May.

Kenny, C. (2003) Why Aren't Countries Rich? Weak States and Bad Neighbourhoods. In Seligson, M. and Passé-Smith, J. (eds) *Development and Under-development: The political economy of global inequality*, 3rd edn. Boulder, CO: Lynne Rienner, pp. 413–25.

Knaus, G. and Martin, F. (2003) Lessons from Bosnia and Herzegovina: travails of the European Raj. *Journal of Democracy* 14(3): 60–74.

Lambert, H. (2015) Yanis Varoufakis – full transcript: Our battle to save Greece. *New Statesman*, 13 July. Available at: http://www.newstatesman.com/world-affairs/2015/07/yanis-varoufakis-full-transcript-our-battle-save-greece (accessed on 8 September 2015).

Larmour, P (2005) *Foreign Flowers: Institutional transfer and good governance in the Pacific Islands*. Honolulu: University of Hawai'i Press.

Leftwich, A. (1996) Two Cheers for Democracy? Democracy and the Developmental State. In Leftwich, A. (ed.) *Democracy and Development*. Cambridge: Polity, pp. 279–95.

Leigh, D. and Evans R. (2010) BAE admits guilt over corrupt arms deals. *Guardian*, 6 February. Available at: http://www.theguardian.com/world/2010/feb/05/bae-systems-arms-deal-corruption (accessed on 8 September 2015).

Mac Ginty, R. (2006) *No War, No Peace: The rejuvenation of stalled peace processes and peace accords*. London: Palgrave.

Mac Ginty, R. (2008) Indigenous peacemaking versus the liberal peace. *Cooperation and Conflict* 43: 139–63.

Mac Ginty, R. and Richmond O. (2007a) Myth or reality: opposing views on the liberal peace and post-war reconstruction. *Global Society* 21(4): 491–7.

MacFarlane N. and Khong, K.F. (2006) *Human Security and the United Nations: A critical history*. Bloomington, IN: Indiana University Press.

Maclean, J. (1988) Marxism and international relations: a strange case of mutual neglect. *Millennium* 17(2): 295–320.

Martz, J. (1996) *The Politics of Clientelism*. New Brunswick, NJ: Transaction.

Mayall, J. and Soares de Oliveira, R. (eds) (2011) *The New Protectorates: International tutelage and the making of liberal states*. London: Hurst.

Migdal, J. (1988) *Strong Societies, Weak States: State–society relations and state capabilities in the third world*. Princeton, NJ: Princeton University Press.

Nasrallah, H. (2008) Speech in solidarity with Gaza, 28 December. Formerly available at: http://almusawwir.org. No longer available.

OECD (2011) New deal for engagement in fragile states. 11 December. Available at: http://www.pbsbdialogue.org/documentupload/49151944.pdf (last accessed on 11 December 2011).

Paris, R. (2004) *At War's End: Building peace after civil conflict*. Cambridge: Cambridge University Press.

Paris, R. (2010) Saving liberal peacebuilding. *Review of International Studies* 36(2): 337–65.

Perlez, J. (2014) US opposing China's answer to World Bank. *New York Times*, 9 October.

Pingeot, L. (2012) *Dangerous Partnership: Private military and security companies and the UN*. New York: Global Policy Forum.

Plant, M. (2014) African Union missing in action in conflicts from Mali to South Sudan. *Guardian*, 6 January.

Pollard, S. (1997) *The International Economy since 1945*. London: Routledge.

Pugh, M. (1997) *The UN, Peace, and Force*. London: Routledge.

Quinn A. and Cox, M. (2007) For better, for worse: how America's foreign policy became wedded to liberal universalism. *Global Society* 21(4): 499–519.

Reno, W. (1997) African weak states and commercial alliances. *African Affairs* 96: 165–85.

Renwick, D. (2015) Peace Operations in Africa. Council on Foreign Relations Backgrounder, 15 May. Available at: http://www.cfr.org/peacekeeping/peace-operations-africa/p9333 (accessed on 8 September 2015).

Richmond, O. (2005a) *The Transformation of Peace*. London: Palgrave.

Richmond, O. (2005b) The Dilemmas of Subcontracting the Liberal Peace. In Richmond, O. and Carey, H. (eds) *Subcontracting Peace: The challenges of NGO peacebuilding*. Aldershot: Ashgate, pp. 19–35.

Richmond, O. and Carey, H. (eds) (2005), *Subcontracting Peace: The challenges of NGO peacebuilding*. Aldershot: Ashgate.

Richmond, O. and Franks, J. (2007) Liberal hubris? Virtual peace in Cambodia. *Security Dialogue* 38(1): 27–48.

Richmond, O. and Mac Ginty, R. (2014) Where now for the critique of the liberal peace? *Cooperation and Conflict* 50(2): 171–89.

Robbins, P. (2003) *Stolen Fruit: The tropical commodities disaster*. London: Zed.

Roberts, D. (2011) *Liberal Peacebuilding and Global Governance: Beyond the metropolis*. London: Routledge.

Sawers, J. (2015) The House of Saud's embryonic embrace of meritocracy. *Financial Times*, 1 May. Available at: http://www.ft.com/cms/s/0/27746568-ef52-11e4-87dc-00144feab7de.html#axzz3gc0Tykz6 (accessed on 8 September 2015).

Seligson, M. (2003) The Dual Gaps: An Overview of Theory and Research. In Seligson, M. and Passé-Smith, J. (eds) *Development and Under-development: The political economy of global inequality*, 3rd edn. Boulder, CO: Lynne Rienner, pp. 1–6.

Shutt, H. (1998) *The trouble with capitalism: An enquiry into the causes of global economic failure*. London: Zed.

Sparrow, A. (2015) Whitehall's King Abdullah half-mast tribute criticised. *Guardian*, 23 January. Available at: http://www.theguardian.com/world/2015/jan/23/king-abdullah-half-mast-flag-tribute-mps (accessed on 8 September 2015).

Spearin, C. (2008) Private, armed and humanitarian? States, NGOs, international private security companies and shifting humanitarianism. *Conflict, Security and Development* 39(4): 363–82.

Stedman, S. (2007) UN transformation in an era of soft balancing. *International Affairs* 83(5): 933–44.

Stiglitz, J. (2003) *Globalization and its Discontents*. New York: Norton.

Stiglitz, J. and Charlton, A. (2005) *Fair Trade for All: How trade can promote development*. Oxford: Oxford University Press.

Sunday Times (2010) Japan's bribes on whaling exposed. 13 June. Available at: http://www.thesundaytimes.co.uk/sto/news/uk_news/Environment/article31661.. ece (accessed on 8 September 2015).

Tadjbakhsh, S. (ed.) (2011) *Rethinking the Liberal Peace: External models and local alternatives*. London: Routledge.

Taylor, I. (2005) *NEPAD: Towards Africa's development or another false start?* Boulder, CO: Lynne Rienner.

The Week (2015) Twelve things women in Saudi Arabia cannot do. 20 July. http://www.theweek.co.uk/60339/twelve-things-women-in-saudi-arabia-cant-do (accessed on 8 September 2015).

Tickner, A. (2001) *Gendering World Politics: Issues and approaches in the post-Cold War era*. Columbia, NY: Columbia University Press.

Tilly, C. (1990) *Coercion, Capital and European States*. Oxford: Blackwell.

Trading Economics (2015) China GDP Annual Growth Rate. Available at: http://www.tradingeconomics.com/china/gdp-growth-annual(accessed on 8 September 2015).

Underhill, G. (2001) The Public Good versus Private Interests and the Global Financial and Monetary System. In Drache, D. (ed.) *The Market or the Public Domain: Global governance and the asymmetry of power*. London: Routledge, pp. 274–95.

United Nations (2014) Letter from President of the General Assembly to President of the Security Council. 14 December. A/69/674-S/2014/911.

United Nations Peacekeeping (2015) Troop and police contributions. Available at: http://www.un.org/en/peacekeeping/resources/statistics/contributors.shtml (accessed on 8 September 2015).

Urquhart, B. and Weiss, T.G. (2012) *What's Wrong with the United Nations and How to Fix It?* Cambridge: Polity.

Wainwright, E. (2003) Responding to state failure – the case of Australia and the Solomon Islands. *Australian Journal of International Affairs* 57(3): 485–93.

Waltz, K. (2001) *Man, the State and War*. Columbia, NY: Columbia University Press.

Whitmeyer, A. (2013) Good news: the US still isn't a failed state. *Foreign Policy*, 2 October. Available at: http://foreignpolicy.com/2013/10/02/good-news-the-united-states-still-isnt-a-failed-state/ (accessed on 8 September 2015).

Williams, A. (1998) *Failed Imagination? New world orders of the twentieth century.* Manchester: Manchester University Press.

Williams, A. (2007) Reconstruction: the bringing of peace and plenty or occult imperialism? *Global Society: Journal of Interdisciplinary International Relations* 21(4): 539–51.

Williams, P. (2011) *War and Conflict in Africa.* Cambridge: Polity.

Woollard, C. (2013) Peace in the post-2015 development goals. *Journal of Peacebuilding and Development* 8(1): 84–9.

World Bank (2011) *World Development Report: Conflict, security and development.* Washington, DC: World Bank.

World Bank (2015) Armed forces personnel, total. Available at: http://data.worldbank.org/indicator/MS.MIL.TOTL.P1 (accessed on 8 September 2015).

Zaum, D. (2006) The authority of international administrations in international society. *Review of International Studies* 32(3): 455–73.

3 People: participation, civil society and gender

Introduction

While the previous chapter examined the institutions and structures that shape conflict and development, this chapter looks at the makers, users, bystanders and victims of these institutions and structures: people. It is interested in the concept and practice of participation, and how particular groups (such as women or civil society) are included or excluded from processes of conflict and development. Public participation is widely regarded as the silver bullet of Western democracy and peacebuilding: through participation comes legitimacy and with legitimacy comes a discourse to justify a particular course of action.

In recent years, 'public participation', 'local participation' and 'local ownership', 'community partnership' and 'customary wisdom' have become prominent themes in relation to the discourses on development and peacebuilding processes (Mac Ginty and Richmond 2013). But to what extent do leading states, international organizations and INGOs deploy these terms as rhetorical devices and to what extent have they a genuine interest in local empowerment in areas that are undergoing peacebuilding and development? This chapter examines the issue of participation in relation to conflict and development. One of the overall themes of this book is a consideration of the extent to which people are included or excluded from peace and development processes. So-called 'ordinary' people are usually excluded from the major decisions that surround war, peace and development, and they are often written out of the histories of these events. The chapter will illustrate how this 'writing out' occurs and consider the steps that have been taken to include or 'write people in' to political and economic processes connected with war and peace.

Participation

As Chapter 2 discussed, many of the structures and institutions connected with conflict and development are elite-led and non-participatory.
In fact, parts of the international financial architecture have been deliberately designed in order to maintain an oligopoly (Glenn 2008). Similarly, many political structures are maintained to limit participation, or
channel it in particular directions. Against a backdrop of the essentially
conservative nature of many political and economic structures, there
has been a growing realization by some international organizations,
governments, INGOs and others that popular participation is the key
to legitimacy and thus sustainability. Interestingly, this realization has
tended to be projected from the Western developed world towards the
developing world, and has not been reflected back into developed
world polities, many of which also suffer from democratic deficits.
Across many development and peacebuilding spheres there has been
an increased emphasis on 'local participation', 'ownership' and
'partnership'.

Plate 7 *An Israeli tank on display at a Hezbollah museum in Lebanon: only a few states
can engage in high-tech war.*

Before examining participation in development and peacebuilding, it is worth noting that popular participation (voluntary and involuntary) is required in conflict (Conteth-Morgan 2004: 112–13). There is a startling disparity between the war-fighting practised by a handful of Western electronically advanced states and that practised by combatants on the rest of the planet. Techno-war as practised by the United States or UK requires relatively few direct protagonists. The actual number of troops deployed on the ground is often quite small due to technology and a division of labour. Even operations in Afghanistan, Iraq or Sierra Leone, in which large numbers of US and UK troops were deployed, benefited from tremendous 'back office' support in the form of communications and coordination bases thousands of miles from the combat theatre (US military operations are directed from Tampa, Florida), while many support functions (for example, catering and logistics) may be conducted by private contractors (Boot 2007). By late 2003 – just months after a full-scale invasion – there were more private security contractors in Iraq than British soldiers (Traynor 2003). Indeed, the combination of professional militaries and 'outsourcing' to private firms means that many citizens in the United States and the United Kingdom are insulated from the fact that their states are engaged in permanent war.

This position differs enormously from most other violent conflict environments. In such cases, the impact and costs of conflict are often felt closer to home and more difficult to avoid. For many people, participation in violent conflict is not a choice and there are few opportunities to escape.

Consider, for example, a family in Syria as the anti-Assad movement escalated into a civil war involving President Assad's regime and a host of anti-regime militant organizations, many of which were engaged in feuding. Family members would have to make tough and risky choices, often based on sketchy information and rapidly changing circumstances. Some family members may be lucky enough to escape to Turkey, Lebanon or Jordan, but this might depend on luck and having access to money. For others, particularly the elderly and infirm (and some women who, for customary reasons, could not travel unaccompanied), escape would be very difficult. Of those left behind, some might depend on the state and outright displays of loyalty to the Assad regime for their survival. Young males conscripted into the military may have to show enthusiasm, while looking for a way out. Others may be able to support, and help, the rebels. But the rebels are a mixed bunch that range from relatively secular opponents of Assad to religious extremists. Again, family members may be faced with uncomfortable choices depending on who held the territory where they lived. The key point is that conflict, war and indeed many development processes, impose invidious decisions on people. Participation in the conflict becomes mandatory because the conflict overwhelms the society and leaves little or no space that can be identified as somehow neutral or beyond the reach of the war.

One of the defining characteristics of the so-called 'new wars' in the post-Cold War period has been the deliberate targeting of civilian populations through 'ethnic cleansing' or the calculated displacement of civilians to cause refugee flows (Kaldor 2012). Conflicts ongoing at the time of writing – Libya, Syria, Ukraine, Iraq, Yemen, South Sudan and Nigeria – have all involved the displacement of large numbers of people. Once displaced, these populations are vulnerable to a variety of problems ranging from public health emergencies to predation by human traffickers. While this is not unique to the post-Cold War period, the displacement of populations often leads to serious knock-on effects in terms of humanitarian emergencies. All organized violence involves 'violence specialists' or actors who take on specific roles during conflicts (Tilly 2002: 20–1). Yet many conflicts involve the mass mobilization of people whether through rallies, the formation of popular movements or recruitment into militant or militant support groups (Horowitz 1985: 443–59). While Western armed forces may be defined by their high-tech methodologies, they also work hard to mobilize populations. The veneration of the US military in the United States, for example, is remarkable. While many US institutions have suffered a decline in legitimacy (the Presidency, a gridlocked Congress, the police, Wall Street, and churches), the military evokes a romanticized loyalty and honour that few dare to criticize. The notion that the military 'can do no wrong' is a powerful mobilizing tool in times of conflict.

Many non-Western military forces (state and non-state) are decidedly informal and low-tech, and rely on large numbers of lightly armed combatants. In some cases, this may involve child soldiers, militias (who farm their land by day, but serve as a 'home guard' by night), 'dollar soldiers' (whose loyalty is bought) or clan-groups (as in Afghanistan) (Marriage 2007). The presence of young unemployed or underemployed males ('a disposable population' in neo-Malthusian parlance) may aid this mass involvement in violent conflict. The chief point is that in such cases war may be a mass-participation activity, rather than an outsourced and compartmentalized activity as it is in Western countries.

The case study of how Sri Lanka decisively won its war with the separatist Liberation Tigers of Tamil Eelam (LTTE) in 2009 is instructive about popular mobilization for conflict. The insurgency had been ongoing since the early 1980s and had racked up high costs including about 100,000 dead, many thousands displaced, the retardation of economic development in Sri Lanka, and the assassination of the Sri Lankan and Indian Presidents. A peace process in the mid-2000s fizzled out, and a new Sri Lankan President in 2005 decided upon an all-out effort to defeat the LTTE. In order to do this, President Rajapaksa invested in preparatory groundwork. Human rights organizations were harassed and booted out of the country (Barnes 2013), loans and military expertise and hardware were sourced from countries with little concern for human rights (Israel, China and Pakistan), government-backed death squads 'disappeared' hundreds of Tamils and some civil society activists so as to minimize voices of opposition (Mac Ginty 2010). But crucially, the Rajapaksa government invested heavily in mobilizing the majority Sinhalese population in favour of a sustained military offensive. This involved much propaganda on behalf of the military and addressing public perceptions of an institution with a mixed reputation. The war against the LTTE became a popular project and, ultimately, a successful one in the sense that the LTTE were defeated.

In addition to actual combatants, mobilizations require political support. Pro- and anti-Syrian forces in Lebanon sought to demonstrate their strength through rallies and counter-rallies in 2006–8. A key form of mobilization comes through political and 'ethnic entrepreneurs' who may seek to mobilize large crowds in support of their cause, often using public demonstrations as a way of indicating their popular support and thus their legitimacy (Kaufman 2001: 5–7). Ukraine, Lebanon, Azerbaijan, Iran, Egypt and Venezuela have seen popular demonstrations and counter-demonstrations, sometimes accompanied by violence and repression.

Political leaders, aided by symbolism and references to a historic past (real or imagined), may be incredibly successful in mobilizing large numbers of people (Mac Ginty 2003: 235–44). War and conflict become, in many cases, mass events involving large numbers of people as participants or active supporters. Over 50,000 people were estimated to have taken part in rioting in Kosovo in March 2004 (United Nations Mission in Kosovo (UNMIK) 2004). Importantly, mass-participation war and conflict may contrast sharply with the situation that prevails during peace negotiations and subsequent peacebuilding in which the opportunities for popular participation are severely constrained. Populations that felt involved in a 'people's war' – as in Colombia or Nepal – may find peace or peace negotiations a much less inclusive process. Box 3.1 illustrates the secrecy that surrounded the origins of the 'Oslo Process' between Palestinians and Israelis.

Box 3.1

The Oslo Peace Process: a closed system

The 1990s 'Oslo Process' between Israel and the Palestine Liberation Organization (PLO) began as clandestine negotiations in a secluded house owned by the Norwegian Ministry of Foreign Affairs on the outskirts of Oslo (Pruitt et al. 1997: 177–82). Staff at the house were told that the occupants were academics working to a strict deadline imposed by an exasperated publisher. At the time of the negotiations, it was illegal for Israeli citizens to meet with members of the PLO. The talks were so secret that many senior members of the PLO and the Israeli government did not know of their existence. The talks were a useful way for each side to test the seriousness of their adversary should more formal and detailed negotiations develop. Problems arose when news of the talks was made public (Wolfsfeld 2003: 92–3). Both the Israeli government and the PLO had invested heavily into demonizing the other side, so they now had to prepare their respective constituencies for the prospect that it might be possible to make a deal with their historic enemies. The seriousness of this task was made clear with the 1995 assassination of one of the architects of the Oslo Process, Israeli Prime Minister Yitzhak Rabin, by a right-wing Israeli who thought that the peace process would lead to ruin. Those involved in the early phase of the Oslo Process faced a real conundrum: do they try to make advances away from the glare and turbulence of publicity, or do they open up the process to public scrutiny before it has had a chance to take root in the minds of the negotiators? Ultimately, the Oslo Process was not sustained. Crucial in its demise was the differential perception of gains and losses held by Israelis and Palestinians. 'Each side to the conflict has continued to see the "other" as being the sole recipient of all the benefits, while not offering anything in return' (Hermann and Newman 2000: 110–11). It is, of course, overly simplistic to say that this perceptual differential was due solely to a failure by each side to prepare their constituencies for the give and take of a peace process, but the interface between the public and private aspects of a peacemaking process requires careful management.

Sources: Pruitt et al. (1997), Hermann and Newman (2000), Wolfsfeld (2003).

The example of peace negotiations in the aftermath of civil war shows how many political networks are 'closed systems' that have firm boundaries of inclusion and exclusion. In the case of peace negotiations, this may be due to understandable security reasons: spoiler groups may be tempted to use violence outside the conference room to influence events inside it (Darby 2001). In practical terms, complex negotiations are often best left to small and dedicated teams who have mastery over their brief. In many cases (for example, peace processes in Sri Lanka, Colombia or Ukraine) the talks have taken place away from the locus of the conflict to allow negotiators to concentrate on wider issues without being blown off course by the immediacy of political violence and the attendant headlines. Problems arise when the existence of talks are made public. Antagonists often invest considerable energy in demonizing their enemies, so to turn around and to admit to having talks with 'the enemy' requires a delicate massaging of the message. Similarly, the contents of the talks and an admission that compromises may have to be made, also require the preparing of constituencies that are more used to being sold messages of complete victory (Darby and Mac Ginty 2003: 267–8).

Opportunities for public involvement in peace negotiations may be limited to public demonstrations to signify a desire for peace (or at least an end to violence), a referendum on a peace accord or new constitution, or the election of a post-peace-accord government (Reilly 2003). Given the circumstances, these opportunities may be valuable (for example, if elections had not been held before or for a very long time), yet they can be best described as one-off events rather than sustained processes that allow for continued and meaningful relationships between citizens and wider political processes. Electoral turnout in some post-conflict and post-authoritarian countries shows that, after initial enthusiasm, many voters realize that elections change little. Post-civil-war parliamentary elections in Mozambique in 1994 had a turnout of 88 per cent. By 2014, the figure was less than 50 per cent. Bosnia-Herzegovina, Cambodia, Timor-Leste, Iraq, Liberia and South Africa have all seen declining voter engagement after initial euphoria (IDEA 2015).

Specialist post-conflict programmes may require or invite the direct participation of particular sectors: for example, members of the security forces may be compelled to take part in a truth and reconciliation process or combatants may be required to disarm and take part in retraining programmes. These formal aspects of a peace process may be described as being overwhelmingly male, limited to specific demographic and political groups, and time-limited. For many other people, however, a peace process may be something that occurs elsewhere (in a capital city) or their only connection with it may be through the media or word of mouth. Thus, the peace process may seem to be 'a creature of the international community and their co-opted national elites and [have] limited connection with the bulk of citizens in the war-affected state' (Darby and Mac Ginty 2008: 5). There may be few opportunities to affect what Harold Saunders (1999) calls 'a public peace process', in which the citizenry can have a substantive input and broaden out formal political processes to involve a wider constituency.

A growing interest in participation among international organizations, INGOs and NGOs is partially a recognition of the dominance of closed systems and the failings that accrue from excluding key constituencies. The pro-participation consensus is based on a belief that people can become 'stakeholders' in projects and feel that as a result of their investment they have ownership in a process. The theory continues that such locally 'owned' processes are more legitimate and more likely to succeed because local constituencies will be able to shape them to suit local needs and aspirations. In such circumstances, external assistance and direction will not be as necessary and local actors will be able to sustain the process. International organizations, INGOs and NGOs now routinely pay attention to issues of participation regarding it as 'a fundamental human right', a way of enhancing 'the sustainability of the settlement' and a means for 'managing inclusion' (ACCORD 2003). Major policy documents from leading international organizations that work on issues of peacebuilding are now full of references to local participation, partnership and legitimacy. Two decades ago that was not the case. The 1992 *Agenda for Peace* document published by the UN is often seen as the seminal explanation of peacebuilding (Boutros-Ghali 1992). The word 'local' is not mentioned in it. By contrast, the World Bank's 2011 *World Development Report* (World Bank 2011) and the 2011 United Nations Development Programme's *Governance for Peace* document (UNDP 2011) mention the word 'local' 382 and 197 times respectively.

There has, however, been criticism that many participatory schemes are superficial and less empowering than their advocates would suggest.

Cooke and Kothari (2002) point to the 'tyranny of participation', whereby communities in developing-world and post-conflict environments are shoehorned into superficial participation mechanisms that suit the requirements of donors (allowing them to tick a box attesting to 'local participation'), but do not necessarily involve local communities in meaningful and sustainable ways. Similarly, Alternative Dispute Resolution techniques have been criticized for their 'coercive harmony' whereby individuals and groups are encouraged to participate but find themselves pressured to reach an accommodation even if it leaves grievances unaddressed and unbalanced power structures in place (Nader 1997). Rather than being truly inclusive, the criticism made in many development and post-conflict environments is that participation mechanisms are just another part of a suite of essentially Western mechanisms that are top–down, and conceived and funded in the West. According to this view, many participation mechanisms are actually part of a process of disempowerment. Different cultural explanations of 'participation' may be in operation. Western notions of participation may prioritize institutional and technocratic mechanisms such as quantifiable, one-off events like elections or consultation exercises. In other societies, participation may be interpreted in more people-centric and relationship-orientated ways, in which participation is a process rather than an event (Lederach 1995: 26). It is important, though, not to romanticize non-Western and indigenous means of popular involvement in political or economic decision-making since many of these processes can reinforce social order and conservatism.

Certain phases of development and peacebuilding processes may be more open to public participation than others. Necessarily, for example, the security and 'stabilization' phases in the aftermath of a violent conflict might actively exclude people and concentrate power in the hands of military forces and a limited number of 'institution builders'. 'Institutionalism before Liberalisation' (Paris 2004) or the belief that stability and functioning bureaucracy must be achieved before democratization has been accepted by most states and international organizations in post-war reconstruction contexts. Yet, at some stage, all development and peacebuilding processes require legitimacy. The ways in which political leaders (international and national) seek to build and maintain legitimacy are crucial. In some societies, guerrilla movements have transformed very quickly into political parties (the morphing of the Kosovo Liberation Army into the Kosovo Democratic Party provides one example). While the transition from protagonist to pragmatist is obviously crucial in any post-civil-war transformation, there is a danger that some political parties are too narrowly based and merely continue the civil war by peaceful means (Gormley-Heenan 2001). The case of Robert Mugabe's Zanu-PF in Zimbabwe provides an example of a party reluctant to leave behind its civil war roots. When under pressure it has repeatedly used the language of liberation and evoked the independence struggle against white rule: a good strategy for mobilizing supporters but of little help for Zimbabwe's economic woes. Many post-civil-war societies have struggled to find a political process that sustains public interest and participation.

Civil society

Civil society can be regarded as both an institution and as a space, and is often regarded as a bulwark against arbitrary control by government. It is defined by the Centre for Civil Society (CCS) as:

> the arena of uncoerced collective action around shared interests, purposes and values. In theory, its institutional forms are distinct from those of the state, family and market, though in practice, the boundaries between state, civil society, family and market are often complex, blurred and negotiated. Civil society commonly embraces a diversity of spaces, actors and institutional forms, varying in their degree of formality, autonomy and power.
>
> (Centre for Civil Society 2008)

Through the media, trade unions, businesses, professional associations, voluntary and church groups, and online campaigns, citizens can become organized, voice their concerns, and engage in activities that may support or challenge a government (Burnell and Calvert 2004). In the Western political mind, an unimpeded civil society is regarded as a key indicator of political freedom. American sociologist Robert Putnam (2000), whose classic study *Bowling Alone* captured the decline of community in the United States, listed the ways in which US citizens could become politically involved:

> contacting local and national officials, working for political parties and other political organizations, discussing politics with our neighbors, attending public meetings, joining in election campaigns, wearing buttons, signing petitions, speaking out on talk radio, and many more.
>
> (Putnam 2000: 31)

Plate 8 *An anti-war protest in Boston: civil society often acts as a bulwark against the state.*

To this we can add the Internet, which has opened up multiple ways through which political views can be aired. For many of us, civil society is non-contentious, has regular and structured access to the corridors of power, and adds colour to the dull routine of politics.

In authoritarian, deeply divided or post-war societies, however, civil society can play a more delicate role and may face significant difficulties in organizing and expressing itself. China and Russia, for example, introduced laws to control civil society and limit the activities of internationally-sponsored NGOs (Denyer 2015). In Israel, those who speak out against the occupation of Palestinian lands and human rights abuses against Palestinians are often targeted by right-wing mobs who seem unhindered by the police (Zonszein 2014). In Cambodia, protestors against land-grabbing (or the transfer of subsistence agricultural land to commercial landowners and agribusinesses) have faced intimidation and have received little protection from the state (Pilorge 2012). General Morsi's counter-revolution in Egypt saw the imprisonment of three Al Jazeera journalists and a clampdown on the media (see Box 3.2). The key point is that civil society is often on the front line in development processes, the prevention of conflict,

---

Box 3.2

Press freedom

By May 2015, the NGO Reporters Without Borders reported that 25 journalists had been killed and 158 had been imprisoned so far that year in Eritrea. While all states and organizations will attempt to influence the media so that they are reflected in a flattering light, some states and organizations attempt to control or suppress the media. The WikiLeaks episode and the harassment and jailing of whistleblowers like Edward Snowden and Bradley Manning showed that Western states can be intolerant of those who publicize the wrongdoings of the state. The World Press Freedom Index, however, shows that the greatest press freedom 'offenders' are developing world states, or authoritarian Arab states like Saudi Arabia and Bahrain. Eritrea props up the Index. The UK government reported that 'The government of Eritrea controls all domestic media outlets and requires all publications to be approved. It is reported that at least 28 journalists are behind bars. Eleven journalists are reported to have been detained since 2001, seven of whom are reported to have died [in captivity]' (FCO 2015). There is no privately owned media and journalists are not permitted to enter the country (Greenslade 2014).

---

and peacebuilding and reconciliation in the aftermath of conflict. Civil society actors are often most prominent in 'bearing witness' or publicizing injustice. However, many actors in conflict and developing contexts make it their business to sideline civil society actors. For example, most peace negotiations include armed groups (state and non-state) but tend to exclude civil society actors. The 2014–15 Havana peace talks held in an attempt to end the long-running civil war in Colombia were unusual in that victims' groups were present at the peace talks.

Recognizing the utility of civil society, international organizations and INGOs often regard it as a key transmission agent for development, conflict prevention and peacebuilding. As a result, immense energy and resources are invested into creating or shoring up civil society. It is here that problems can emerge. Western interveners may have a particular concept of civil society in mind that may not always fit with the political, social and cultural norms of the host society. Western actors may attempt to create civil society in the image of their home civil society, oblivious to the fact that a vibrant civil society exists in the post-war or developing world context. A classic example of this comes in the form of the US response to the looting that followed the 2003 Anglo-American invasion of Iraq (Mac Ginty 2004). According to US military commanders, the problem was that there was no civil society that could control the civilian population and coordinate attempts to repatriate looted goods with their original owners (BBC News 2003). The view that Iraq was bereft of civil society was widespread. According to Toby Dodge, 'Before the liberation of Baghdad it was impossible to talk about civil society in Iraq … autonomous collective societal structures beyond the control of the Ba'athist state did not survive' (cited in Pirouz and Nautré 2005: 3–4). Yet, notwithstanding the brutalities of the Saddam Hussein regime, there was some sort of civil society in place: mosques, family groups, professional associations, local Red Crescent societies and other networks. Certainly these networks were very far removed from the plural and voluntary forms of civil society found in Western liberal democracies. Yet, to assert that there is 'no civil society' may over-look various forms of social capital that may not be immediately visible to Western eyes.

In many post-war societies, Western interveners have taken it upon themselves to create civil society, often in the image of Western models (Prusher 2003). Through 'local capacity building' schemes and the direct funding of NGOs, international actors can create a civil society that is recognizable to them, but may not necessarily have deep and effective roots in the host society (Adamson 2002: 200). One may be forced to ask: what is the essential purpose of this new civil society? Its key role is often to provide a recognizable interface with which international organizations and INGOs can engage. The connection between civil society and the host society may be thin, limited to a particular strata of the host country that has been artifi-cially created by international actors. The new civil society may be unsustainable because of its dependence on Western funding streams. A Western-created and -inspired civil society, or the 'democracy sector', is often highly visible in post-war societies. It is often urban, metropolitan and English-speaking (Mac Ginty 2006: 52). Indeed, in many cities, it often inhabits the same physical space as the INGO sector that was responsible for spawning it, and fuels the same micro-economy of vehicle hire, Western-style coffee shops and bars, and a conference support sector. The key issues with civil society in any context are the extent to which it can be inclusive, voluntary and act as a real counterweight to state institutions. These issues are particularly acute in deeply divided societies in which society may be structured in a way that excludes or discriminates against particular groups. Elaborate structures (for example, human rights commissions) or legal codes (that outlaw discrimination) may be created in the wake of a civil war, but it does not always follow that institutional engineering will bring with it changes in societal behaviour and thought. India has had legislation outlawing the caste system for decades, but that system of apartheid is still firmly rooted in everyday life.

It is worth stressing that the 'creation' of civil society by external agents and consequent problems between the new civil society and the host society are by no means restricted to post-conflict settings. The phenomenon of the artificial civil society, especially the democracy-promotion sector, is common in many non-conflict developing contexts.

Importantly, civil society need not be restricted to one polity or location. Civil societies can be transnational, especially so in a globalized and digitally connected age. Indeed, given that so many conflict and development-related problems are transnational, it makes sense that civil society mirrors this. Moreover, some locations are so unwelcoming to dissenting civil society organizations, that the only meaningful opposition can come from outside the state boundaries in the form of diaspora organizations.

The concluding point of this section is to differentiate between civil society as a set of organizations and civil society as a behavior. There is a danger that civil society is reduced to being civil society organizations such as NGOs with offices, staff, programmes and projects. This type of civil society (while being valuable in many cases) risks being shallow. Its agendas might be donor led, it exists only so long as overseas bodies fund programmes, and it may be incestuously urban with the same faces turning up to a narrow range of events in the same city and saying the same thing. Travel to Belfast, Sarajevo, Pristina, Beirut and other 'post-conflict cities' and one is likely to realize that there is a small world of civil society actors who speak the same professional language, attend the same events and often socialize together. A broader view of civil society can be more fruitful, however. This view regards civil society as a phenomenon not just restricted to civil society organizations. Instead, it regards civil society as a much more organic and evolutionary space comprised of individuals, groups and organizations. This space will be a complex and dynamic tapestry of relations. As befits a deeply divided society, it may be riven with fractures and reflect major sectarian or identity divisions. Much civil society operates at the intra-group level. Here it may be divisive and contribute to the intensification of division in society. But, alongside this negative purpose, it might also be positive at the intra-group level: providing social services, restraint and guidance.

In this view, civil society is complex and dynamic. It can simultaneously restrain and inflame conflict. It is not restricted to artificially created organizations, and instead inhabits the realm of everyday life. It is made and remade on a daily basis through intra- and inter-group contact. It occurs in the stairwells of apartment buildings, on the street in the city neighbourhood and in the village market. This everyday civil society might be difficult for outsiders to spot, but it can contribute to the glue that prevents a society from moving from chronic tension to all-out war (Mac Ginty 2014).

Gender

The 1997 United Nations Human Development Report noted starkly that 'no society treats its women as well as its men' (UNDP 1997: 39). There is every reason to believe that this judgement still holds true. Even in apparently fair and peaceful societies, the gap between male and female experiences is startling. According to 2015 figures, in the UK, despite legislation banning discrimination, women – on average – earn about 80 per cent of what men earn (Topping 2015). There are over three times as many men Members of Parliament as women. Only 8.6 per cent of FTSE 100 companies have female executive directors (Smith and Rigby 2015). These gender differentials may be stretched and distorted during conflict and development.

Discussion of gender issues in relation to conflict is made difficult by a potential minefield of stereotyping in which women are characterized as life-givers, nurturers and generally pacific, while men are depicted in the warrior mode (Jacoby 2008: 92). Such stereotypes are best avoided. There is also a tendency in some literature and practice to lump women together as 'victims' or an undifferentiated category without agency. Often 'womenandchildren' can be found as an apparent afterthought in the later stages of official reports. Having opened with a caveat, we can then say that development, conflict and peace impact on men and women in different ways, with the burdens and benefits shared unevenly. But we cannot make blanket statements that have universal applicability; clearly circumstances and context matter.

The principal gender distinction between men and women in times of conflict is that men are often (but not always) the main combatants, while women often perform (but not always) ancillary or support roles (Turshen and Twagiramariya 1998; Rehn and Johnson Sirleaf 2002). But as has been pointed out frequently in this book, contemporary civil wars and contexts of political suppression are very far removed from the classical notion of warfare with professional militaries and clear lines of distinction between combatants and non-combatants. This blurring of the demarcation between combatants and non-combatants has implications for how war impacts on men and women. Often entire ethnonational groups are targeted or mobilized, regardless of gender. Within these groups men and women may be affected in different ways (women more prone to sexual violence or men more prone to arrest and torture) but the group may suffer the same overall experience. Women may not be the main frontline targets of groups like Islamic State or Boko Haram, but once these organizations seize territory, then women become the object of social policing.

Development processes and the ending of wars provide 'transition moments' in which it is possible to make significant societal changes, for example in relation to land reform, public access to power, or the treatment of women. It is not always clear that these opportunities are taken. Certainly in peace negotiations, issues of territory, security and constitutional design are often prioritized before 'secondary' issues such as gender, truth recovery or social inclusion. If, at best, peace processes can be labelled as 'gender blind', then they bring with them the danger that they reinforce the existing patriarchy. It may be the case that external actors, such as foreign governments and INGOs, favour the inclusion of gender reform in peace negotiations or the formulation of a new constitution. Local actors may not agree: either they may not see the issue as urgent or they may see it as undesirable as it confronts dominant cultural norms. Or local civil society may advocate gender reforms that mark a significant advance for the country in question but are not as radical as some Western actors would favour. This raises an interesting series of questions: whose definition of gender equality is to be followed? What are the main components of any gender equality programme? Will the promotion of gender equality place women in danger? Such 'clashes of culture', between Western norms and indigenous practices, are common in virtually every aspect of development and post-conflict reconstruction. It is important to note that 'Western' and 'indigenous' do not comprise discrete categories: they vary enormously and all social organizations and practices are – to some extent – hybrids. But the key struggle between Western and indigenous norms persists. It manifests itself in Western notions that there are 'proper' forms of social organization. Conflict may arise as different elements in developing and post-conflict societies may attempt to resist, accept or subvert external development interventions and the cultural baggage associated with them.

It is one thing to outlaw gender discrimination, but it is another thing to change the cultural norms and thought processes that support it. As Antonia Potter (2008: 105) notes, 'reality lags far behind rhetoric on women's involvement in peace processes.' Certainly an impressive legal and normative framework has been erected internationally and in many national contexts over the past few decades to protect the rights of women and encourage their full participation in political life. The 1979 Convention on the Elimination of All Forms of Discrimination against Women, the 1995 Beijing Declaration and Platform of Action and the 2000 UN Security Council Resolution 1325 on Women, Peace and Security, all count as international landmark documents for the inclusion of women. These, and other initiatives, are testament to the incredible energy of women's movements that often operate in unwelcoming environments. Crouch (2004) calls women's mobilization 'a great democratic phenomenon' that has included

> extreme radicals; sober reformist policy-makers; cunning reactionaries taking the movement's messages and reinterpreting them; both elite and popular cultural manifestations of many kinds; the gradual suffusion into the conversation of ordinary people of elements of the language of an initially esoteric movement.
>
> (Crouch 2004: 62)

Yet, as Potter (2008) politely observes,

> it might not be unreasonable to worry that the slow pace of change implies that misogynist or bigoted views continue to hold a certain amount of sway in the most progressive and liberal societies, even if unconsciously … A review of the literature and case studies on post-conflict situations, reveals … a depressing paucity of examples of implementation.
>
> (Potter 2008: 107)

The very fact that global activism on gender issues is organized around 'Beijing +20' (that is, the Beijing Summit 20 years on) suggests that many of the original goals of equality have not been met. It is sobering to think of the gender differentials that exist despite the enormous political, economic and social change experienced by many societies since the 1950s and the massive development and peace-support interventions since the end of the Cold War. Janet Hunt (2004) sums up the situation thus:

> Women are said to be 70 per cent of the world's poor; have only about 10 per cent of all parliamentary seats in most countries; and in developing countries, on average, they earn only 73 per cent of male earnings. There is not a single developing country in which women and men enjoy equal rights under the law. In particular, women are discriminated against in areas such as their right to land and property, and their right to conduct business independently. The result is that women are more vulnerable to poverty than men, especially as a result of widowhood, separation or divorce, and the consequent loss of access to productive assets.
>
> (Hunt 2004: 243)

Many development interventions have sought ways in which to include women in economic growth. Thus women's participation is built into many development projects or programmes (see Box 3.3 on micro-finance). The hope is that multiple small schemes will have a cumulative impact, not only on the lives of women but also on the societal and structural factors that dictate women's position in society.

A final issue to consider in relation to gender and conflict is the role of equality. For some, full equality means the ability for females to serve in frontline positions. Thus, for example, some news sources have trumpeted Pakistan's first female fighter pilot as 'a role model for millions of girls' (Crilly 2013). Yet, much feminist literature would contest the notion that the equality to kill is equality at all. For critical feminists, it is merely the acceptance of a female into a male-dominated structure and does not alter the basic male dominance of systems of government and militarism (Enloe 2000, 2007; Sjoberg 2013). Most feminist theorists prioritize structural change of the system over piecemeal 'equality' that implicates women in a system that is unjust, violent and fundamentally exploitative of women.

Box 3.3

Women's empowerment through micro-finance initiatives

Many micro-credit or micro-finance initiatives have had a deliberate gender dimension. Indeed, in 2006 seven out of ten micro-finance clients were believed to be women (Armendáriz and Broome 2008). Small loans have been targeted at women in the hope that the resulting small businesses will first help alleviate poverty for them and their families, and second act as a means of empowerment. Women have been given loans to establish themselves as small-scale chicken farmers or to come together as a cooperative to market their handicrafts (Bigsten 2003: 115). Mayoux (2002) notes how

> Microfinance programmes have significant potential for contributing to women's economic, social and political empowerment. Access to savings and credit can initiate or strengthen a series of interlinked and mutually reinforcing 'virtual spirals' of empowerment. Women can use savings and credit for economic activity, thus increasing incomes and assets and control over income and assets.
>
> (Mayoux 2002: 76)

There has been something of a bandwagon championing micro-credit. The UN declared 2005 the 'Year of Micro-credit' and the Grameen Bank (a Bangladeshi initiative that has subsequently spread to 28 countries) was awarded a Nobel Prize. There is no doubting that micro-credit schemes have economically empowered some women, and there is evidence that there have been additional one-off benefits in terms of social capital and public perceptions of women (Sanyal 2006). Critics, however, argue that not only are the economic benefits of micro-credit schemes often overblown, but also the schemes may not be as liberating as their advocates suggest. For example, it is suggested that if economic empowerment is not linked to other empowerment (for example, greater legal powers for women such as the right to buy and sell land in certain countries) then stand-alone micro-credit schemes will have limited impact (SIDA (Swedish International Development Cooperation Agency) 2006). Others have noted that indebtedness loads women with an additional burden and may bring them into conflict with men in the home. Just as importantly, without structural change to the organization of the economy (almost impossible given the power of international economic forces), any changes linked with micro-credit schemes or pro-poor growth strategies are likely to be sporadic and dependent on local circumstances.

Sources: Mayoux (2002), Bigsten (2003), Sanyal (2006), SIDA (2006), Armendáriz and Broome (2008).

Conclusions

A fundamental issue running through this book is culture and the potential of Western norms and practices connected to conflict, peace-building and development to clash with local, indigenous or traditional norms and practices. We should, of course, be careful not to build a strict dichotomy in which everything from 'the West' (by no means a discrete category) can be described as top–down, legalistic and rational, and everything found *in situ* in developing and post-conflict environments as bottom–up, organic and traditional. There is a very real danger of romanticizing the local and the traditional. At heart is the question of the appropriateness of external intervention and the appropriateness of the type of intervention. A case-by-case approach is best. In some cases, external states and actors are the only ones with the capability to effect the scale of change required to make a difference (for example, inoculation campaigns by Médecins Sans Frontières). In other cases, the actions of external (often Western) states or actors undermine the capability of local states and actors. Many development and peacebuilding contexts are the scene of a cultural conflict between what might be described as variations of Western liberalism, and local belief systems and practices of social and political organization. These conflicts revolve around the extent to which liberalism is adopted, enforced or resisted.

It is worth ending a chapter on participation by restating the agency of local populations. It is tempting to depict local communities as passive actors who receive or consume conflict, peace and development that is made elsewhere. Certainly the international and transnational nature of conflict and development means that there are few hermetically sealed contexts on the planet. All societies are penetrated by external forces to some degree, and the sheer scale of development and peace-support interventions is awesome in some contexts. But we should not overlook the local. Keesing (1992: 2) stresses the importance of seeing 'peripheral populations as active agents in shaping and controlling their engagement with the outside world, giving local meaning to alien ideas, institutions and things, in various ways resisting them'. Local communities have indeed power to absorb, renegotiate, subvert and resist external pressure. Clearly, this power will differ from context to context, but one of the most remarkable aspects of globalization and liberal peace interventions is the extent of local variation. If nothing else, this local variation is a sign of participation in development, conflict and pacific processes.

Every international intervention in another country has unanticipated consequences. Whether it is the US invasion of Iraq, the Saudi Arabian bombing of Yemen, or the presence of Nigerian peacekeepers in Darfur, local people react in ways that military and humanitarian planners did not foresee. The more farsighted of the international interveners are able to find ways to work with local actors, to calibrate their intervention, and to know when to leave.

Summary

- Different types of conflict, peacebuilding and development have different ways of excluding and including people.
- Peacebuilders and those engaged in development processes often struggle to establish wide levels of popular participation in their projects.
- Civil society is often regarded as an important bulwark against the unrestrained actions of governments or corporations, but there is a danger that Western notions of civil society are regarded as best.
- It is important to see civil society as more than just civil society organizations. It is also a space that can be organic and evolutionary.
- Development and peacebuilding processes consistently sideline women.
- There is a serious clash between Western notions of gender rights and notions of gender rights in non-Western contexts.
- At the centre of many conflicts and tensions between Western and non-Western actors is culture: an issue that Western observers often overlook.

Discussion questions

- How should development projects attempt to gain popular legitimacy and acceptance in the context of a developing country?
- What avenues of political participation are available to you in your own country? Would you recommend them to people in other countries?
- Must civil society be open to all groups, especially in a deeply divided society?
- Should women in Afghanistan enjoy the same rights as women in America?

Further reading

There is an extensive literature on the role of civil society in peace-building, with Thania Paffenholz's (2012) *Civil Society and Peace-building: A critical assessment*, Boulder, CO: Lynne Rienner, probably the most comprehensive and relevant. Issues of culture (and how we interpret other cultures) are very important to inclusion and exclusion in development and conflict. Good starting points are P. Chabal and J.-P. Daloz (2006) *Culture Troubles: Politics and the interpretation of meaning*, London: Hurst, P. Chabal (2012) *The End of Conceit: Western rationality after postcolonialism*, London: Zed and Pankaj Mishra (2012) *From the Ruins of Empire: The revolt against the west and the remaking of Asia*, London: Allen Lane. Three modern classics are also worth reading to help us reflect about culture ('theirs' and 'ours'): S. Huntington, The clash of civilizations?, in *Foreign Affairs* 1993, 72(3); R. D. Kaplan, The coming anarchy, *Atlantic Monthly*, February 1994; and E. Said (1979) *Orientalism*, London: Vintage. Anthropology, journalism and personal testimony often give the best perspectives on the impact of conflict on women. See, for example, S. Drakulic (1994) *Balkan Express: Fragments from the other side of war*, London: Perennial or Halima Bashir (2012) *Tears in the Desert: One woman's true story of surviving the war in Darfur*, London: Hodder and Stoughton.

Useful websites

Many international organizations and international non-governmental organizations have information on their websites labelled 'participation' or 'local partners', though it is useful to consider where the power lies in relationships between local and international actors. See, for example, the Asian Development Bank (www.adb.org) or ECHO – the European Commission's Humanitarian Aid Office (ec.europa.eu/echo/). United Nations Women (http://www.unwomen.org/en) contains very useful information on gender, development and conflict. Other development-related organizations, such as the World Bank (www.worldbank.org) or USAID (www.usaid.gov), contain material on the gender dimension of their programmes. You may have to use the search function on these organizations' websites. See also the gender-related information maintained by Human Rights Watch (www.hrw.org), the Hunt Alternatives Fund (www.huntalternatives.org), International Alert (www.international-alert.org) and Amnesty International (www.amnesty.org).

References

ACCORD (2003) Owning the Process: Public participation in peacemaking – principles to guide policy and practice. Presentation by Conciliation Resources ACCORD

Programme at an International Peace Academy Conference. New York. 12 February. Available at: http://www.c-r.org/downloads/Owning%20the%20process.pdf (accessed 18 March 2008).

Adamson, F. (2002) International Democracy Assistance in Uzbekistan and Kyrgyztan. In Mendelson, S. and Glenn, J. (eds) *The Power and Limits of NGOs: A critical look at building democracy in Eastern Europe and Eurasia*. New York: Columbia University Press, pp. 177–206.

Armendáriz B. and Broome, N. (2008) *Gender Empowerment in Microfinance*. Cambridge, MA: Harvard University Press.

Barnes, J. (2013) Making torture possible: the Sri Lankan conflict 2006–2009. *Journal of South Asian Development* 8(3): 333–58.

BBC News (2003) Baghdad protests over looting. *BBC News Online*. 12 April. Available at: http://news.bbc.co.uk/1/hi/world/middle_east/2941733.stm (accessed on 19 March 2008).

Bigsten, A. (2003) Selecting Priorities for Poverty Reduction and Human Development in Ethiopia. In Addison, T. (ed.) *From Conflict to Recovery in Africa*. Oxford: Oxford University Press, pp. 106–22.

Boot, M. (2007) In defence of Blackwater. *The Press* (Christchurch). 10 October.

Boutros-Ghali, B. (1992) *An Agenda for Peace, Preventive diplomacy, peacemaking and peace-keeping*. New York: United Nations.

Burnell, P. and Calvert, P. (eds) (2004) *Civil Society in Democratisation*. London: Frank Cass.

Centre for Civil Society (2008) *What Is Civil Society?* CCS website. Available at: http://www.lse.ac.uk/collections/CCS/what_is_civil_society.htm (accessed on 24 July 2015).

Conteth-Morgan, E. (2004) *Collective Political Violence: An introduction to the theories and cases of violent conflicts*. New York: Routledge.

Cooke, W. and Kothari, U. (eds) (2002) *Participation: The new tyranny?* London: Zed.

Crilly, R. (2013) Pakistan's only female fighter pilot becomes role model for millions of girls. *Daily Telegraph*, 1 September. Available at: http://www.telegraph.co.uk/news/worldnews/asia/pakistan/10279119/Pakistans-only-female-fighter-pilot-becomes-role-model-for-millions-of-girls.html (accessed on 8 September 2015).

Crouch, C. (2004) *Post-democracy*. Cambridge: Polity.

Darby, J. (2001) *The Effects of Violence on Peace Processes*. Washington, DC: United States Institute of Peace.

Darby, J. and Mac Ginty, R. (2003) Conclusion: Peace Processes, Present and Future. In Darby, J. and Mac Ginty, R. (eds) *Contemporary Peacemaking: Conflict, violence and peace processes*. Basingstoke: Palgrave, pp. 256–74.

Darby, J. and Mac Ginty, R. (2008) Introduction: What Peace? What Process? In Darby, J. and Mac Ginty, R. (eds) *Contemporary Peacemaking: Conflict, peace processes and post-war reconstruction*, 2nd edn. Basingstoke: Palgrave, pp. 1–9.

Denyer, S. (2015) NGOs in China fear clampdown. *Guardian*, 30 March.

Enloe, C. (2000) *Maneuvers: The international politics of militarizing women's lives*. Berkeley, CA: University of California Press.

Enloe, C. (2007) *Globalization and Militarism: Feminists make the link*. New York: Rowman and Littlefield.

Foreign and Commonwealth Office (2015) Eritrea: Country of Concern. FCO, 12 March. Available at: https://www.gov.uk/government/publications/eritrea-country-of-concern--2/eritrea-country-of-concern (accessed on 10 May 2015).

Glenn, J. (2008) Global governance and the democratic deficit: stifling the voice of the South. *Third World Quarterly* 29(2): 217–38.

Gormley-Heenan, C. (2001) From Protagonist to Pragmatist: Political leadership in societies in transition. Report published by INCORE, University of Ulster/United Nations University.

Greenslade, R. (2014) Press freedom body highlights plight of Eritrea's jailed journalists. *Guardian*, 8 December.

Hermann, T. and Newman, D. (2000) A Path Strewn with Thorns: Along the Difficult Road of Israeli–Palestinian Peacemaking. In Darby, J. and Mac Ginty, R. (eds) *The Management of Peace Processes*. Basingstoke: Macmillan, pp. 107–53.

Horowitz, D. (1985) *Ethnic Groups in Conflict*. Berkley, CA: University of California Press.

Hunt, J. (2004) Gender and Development. In Kingsbury, D., Remenyi, J., McKay, J. and Hunt, J., *Key Issues in Development*. Basingstoke: Palgrave Macmillan, pp. 242–65.

IDEA (2015) Voter Turnout. International IDEA website. Available at: http://www.idea.int/vt/ (accessed on 3 June 2015).

Jacoby, T. (2008) *Understanding Conflict and Violence: Theoretical and interdisciplinary approaches*. London: Routledge.

Kaldor, M. (2012) *New and Old Wars: Organized violence in a global era*, 2nd and 3rd edn. Cambridge: Polity.

Kaufman, S.J. (2001) *Modern Hatreds: The symbolic politics of ethnic war*. Ithaca, NY: Cornell University Press.

Keesing, R. (1992) *Custom and Confrontation: The Kwaio struggle for cultural autonomy*. Chicago, IL: University of Chicago Press.

Lederach, J.P. (1995) *Preparing for Peace: Conflict transformation across cultures*. Syracuse, NY: Syracuse University Press.

Mac Ginty, R. (2003) The Role of Symbols in Peacemaking. In Darby, J. and Mac Ginty, R. (eds) *Contemporary Peacemaking: Conflict, violence and peace processes*. Basingstoke: Palgrave, pp. 235–44.

Mac Ginty, R. (2004) Looting in the context of violent conflict: a conceptualization and typology. *Third World Quarterly* 25(5): 857–70.

Mac Ginty, R. (2006) *No War, No Peace: The rejuvenation of stalled peace processes and peace accords*. London: Palgrave.

Mac Ginty, R. (2010) Social network analysis and counterinsurgency: a counterproductive strategy? *Critical Studies on Terrorism* 3(2): 209–27.

Mac Ginty, R. (2014) Everyday peace: bottom-up and local agency in conflict-affected societies. *Security Dialogue* 45(6): 548–64.

Mac Ginty, R. and Richmond, O. (2013) The local turn in peace building: a critical agenda for peace. *Third World Quarterly* 34(5): 763–83.

Marriage, Z. (2007) Flip-flop rebel, dollar soldier: demobilisation in the Democratic Republic of Congo. *Conflict, Security and Development* 7(2): 281–309.

Mayoux, L. (2002) Microfinance and women's empowerment: rethinking best practice. *Development Bulletin* 57: 76–81.

Nader, L. (1997) Controlling processes: tracing the dynamic components of power. *Current Anthropology* 38(5): 711–38.

Paris, R. (2004) *At War's End: Building peace after civil conflict.* Cambridge: Cambridge University Press.

Pilorge, N. (2012) Conflict over land in Cambodia is taking a dangerous turn. *Guardian*, 25 September.

Pirouz, R. and Nautré, Z. (2005) *An Action Plan for Iraq: The perspective of Iraqi civil society.* London: Foreign Policy Centre Report.

Potter, A. (2008) Women, Gender and Peacemaking in Civil Wars. In Darby, J. and Mac Ginty, R. (eds) *Contemporary Peacemaking: Conflict, peace processes and post-war reconstruction,* 2nd edn. Basingstoke: Palgrave, pp. 105–19.

Pruitt, D.G., Bercovitch, J., Zartman, I.W. (1997) A brief history of the Oslo talks. *International Negotiation* 2(2): 177–82.

Prusher, I. (2003) Iraq's new challenge: civil society. *Christian Science Monitor.* 8 October.

Putnam, R. (2000) *Bowling Alone: The collapse and revival of American community.* New York: Simon & Schuster.

Rehn, E. and Johnson Sirleaf, E. (2002) *Women, War, Peace: The independent experts' assessment of the impact of armed conflict on women and women's role in peace building.* New York: UNIFEM.

Reilly, B. (2003) Democratic Validation. In Darby, J. and Mac Ginty, R. (eds) *Contemporary Peacemaking: Conflict, violence and peace processes.* Basingstoke: Palgrave, pp. 174–83.

Sanyal, P. (2006) *Credit, Capital and Collective Action: Microfinance and pathways to women's empowerment.* Paper presented at the annual meeting of the American Sociological Association, Montreal Convention Center, Montreal, Quebec, Canada.

Saunders, H. (1999) *A Public Peace Process: Sustained dialogue to transform racial and ethnic conflict.* Basingstoke: Palgrave.

SIDA (2006) *Microfinance and Women's Empowerment: Evidence from the self help group bank linkage programme in India.* Stockholm: SIDA.

Sjoberg, L. (2013) *Gendering Global Conflict: Towards a feminist theory of conflict.* Columbia, NY: Columbia University Press.

Smith, A. and Rigby, E. (2015) Women make inroads on UK boards but top slots remain elusive. *Financial Times*, 15 March. Available at: http://www.ft.com/cms/s/0/ab49c1e4-d215-11e4-b66d-00144feab7de.html#axzz3gPw1vtqQ (accessed on 8 September 2015).

Tilly, C. (2002) Violent and Non-violent Trajectories in Contentious Politics. In Ungar, M., Bermanzohn, S. and Worcester, K. (eds) *Violence and Politics: Globalization's paradox.* New York: Routledge, pp. 13–31.

Topping, A. (2015) Gender pay gap will not close for 70 years at current rate, says UN. *Guardian*, 5 March. Available at: http://www.theguardian.com/money/2015/mar/05/gender-pay-gap-remain-70-years-un (accessed on 8 September 2015).

Traynor, I. (2003) The privatisation of war. *Observer*, 10 December.

Turshen M. and Twagiramariya, C. (1998) *What Women Do in Wartime: Gender and conflict in Africa.* London: Zed.

United Nations Development Programme (1997) *Human Development Report 1997: Human development to eradicate poverty.* New York: United Nations.

United Nations Development Programme (2011) *Governance for Peace: Securing the social contract.* New York: UNDP.

UNMIK (2004) Condemning violence in Kosovo, Security Council demands return to rule of law. UNMIK Online. 18 March. Available at: www.unmikonline.org (accessed on 18 March 2008).

Wolfsfeld, G. (2003) The Role of the News Media in Peace Negotiations: Variations Over Time and Circumstance. In Darby, J. and Mac Ginty, R. (eds) *Contemporary Peacemaking: Conflict, violence and peace processes.* Basingstoke: Palgrave, pp. 87–99.

World Bank (2011) G7+ A paradigm shift for fragile states. World Bank Institute, 13 December. Available at: http://wbi.worldbank.org/wbi/stories/g7-paradigm-shift (accessed on 8 September 2015).

Zonszein, M. (2014) How Israel silences dissent. *New York Times*, 26 September.

4 ▸ Conflict transformation and development

Introduction

The next three chapters are essentially about attempts to effect 'transitions' from conflict and war to a kind of 'peace'. To put it in the words of the OECD's *Handbook on Security Sector Reform* (OECD 2007: Foreword),

> Recent debate within the international community has centred on the challenge of insecurity and conflict as a barrier to political, economic and social development. If states are to create the conditions in which they can escape from a downward spiral where insecurity, criminalization and under-development are mutually reinforcing, socio-economic and security dimensions must be tackled simultaneously. The traditional concept of security is being redefined to include not only state stability and the security of nations, but also a clear focus on the safety and well-being of their people.
>
> (OECD 2007)

These three chapters collectively will first look at the main strategies employed by the international community or individual powers or groups to try and do something about the conflicts and wars in developing (and some more developed) countries of which we have so far described the genesis. This can be by way of gentle or more muscled intervention, often referred to generically as 'peacebuilding'. Roland Paris, the author of a very influential book (Paris 2004) on the subject, describes it as follows:

> I define peace building as an activity that takes place in a post-civil war environment, the purpose of which is to create the conditions for a stable and lasting peace and to prevent the recurrence of large-scale violence... Since the late 1990s, the term 'peace building' has been used to describe a much broader range of local, regional, national, and international initiatives intended to promote peace, including conflict prevention.
>
> (Paris 2004)

As Vivienne Jabri puts it, this catch-all phrase has become a 'norm', and one 'that informs the normative underpinning and legitimization of particular kinds of intervention'. It is nearly always supervised and implemented by the UN or another IGO, like NATO (Jabri 2013: 3–4).

As Paris indicates, there has been a revival of interest in the idea of such peacebuilding needing very 'local' agents to make it work, or in David Chandler's words, a 'discovery of the local' (Chandler 2013: 22). Underpinning nearly all such attempts is the idea that they should aim to create societies that are liberal economically and politically. So, second, these chapters will investigate how successful such strategies have or have not been; and, third, they will ask why such strategies have failed or succeeded and how we might try other tactics, up to and including the possibility of total non-intervention.

In all these considerations we have to take into account that the overarching categories we use are shorthand for much more complex and integrated issue areas. Reconstruction, the delivery of aid, or other peacebuilding attempts cannot be separated from attempts at conflict resolution or 'reconciliation' after wars. They may take place sequentially, or simultaneously, while the violence continues at varying levels of intensity (see the debate on 'contingency' in Keashley and Fisher 1990). Equally, each of these concepts is a *matrioshka* doll of other concepts – disarmament, demobilization and reintegration (DDR) of ex-combatants being one important embedded 'doll' that will be explored in Chapter 5.

This chapter will first look at the concepts that can be used for understanding the practices of conflict resolution (or as we prefer, 'transformation'), as most famously epitomized by attempts to bring about peace between Israel and Palestine in the 1990s. It has to be said that the hope associated with such agreements as the Oslo Accords of 1993 has faded and they are generally now seen as a failure, replaced by a feeling that such conflicts are 'intractable', especially in the light of the religious conflicts that now seem endemic in the area after the 'Arab Spring' (Fox 2013; Mitchell 2014). This will include a discussion of the differences between 'bottom–up' and 'top–down' approaches to conflict settlement, management and resolution/transformation as they are discussed in the mainstream literature.

It will, second, look at other techniques and ideas that have been developed to try and heal communities after wars, and in particular techniques of 'reconciliation' and 'retributive justice' through truth and reconciliation commissions and war crimes tribunals. The over-arching philosophical questions that underlie this inquiry and that we will engage with will be: how can we arrive at a shared understanding of the past, and of 'truth', of 'history' indeed; and how can we envis-age the 'reintegration' of a society damaged by war, or as Rigby puts it, the 'envisioning of a common future together' (Rigby 2001: 12)?

Dealing with conflict – from 'settlement' to 'sustainability'

The period since the end of the Cold War has seen a swing towards optimism about the possibilities of dealing with conflict in a satisfac-tory way from third party intervention of various kinds through to deep pessimism in the 2000s as 'intractability' has seemingly increased. We could even now speak of a 'nostalgia' for the 'conflict resolution' hopes of the 1990s. Certainly the ideal of 'an equal recog-nition of the stakeholders involved' (Jabri 2013: 16), which in Afghanistan, for example, might mean giving equal weight to the views of the Taliban and the Afghan government, has been replaced by a belief in the necessity of the use of force in the service of the 'international community', by which it is usually meant the UN and powerful interventionary states.

Equally, and as we have noted, the nature of war has certainly changed, from the predominantly international wars of the Cold War to the intra-state ones after it. In the 1990s there were approximately five intra-state conflicts for every one international war. The think tank PIOOM reported that the 1990s had one of the lowest numbers of battle deaths of the twentieth century, ('World Conflict and Human Rights Map' by PIOOM, atf_world_conf_map.pdf). This was not-withstanding the highly publicized wars in the Former Yugoslavia (1991–9), where an authoritative figure of 130,000 deaths has been recorded (Humanitarian Law Center). The outbreak of civil wars in the Middle East and Eastern Europe since the 'Arab Spring' of 2011 and the events in Ukraine since 2014 has seen a very disturbing upward trend in the most recent 2014 figures for deaths in armed conflicts (175,881, up 62,975 from 2013 (International Institute for Strategic Studies 2015a)).

There are many reasons not to feel complacent about these figures. In the first place the nature of warfare has been changing since the 1990s, both in what is often referred to as 'hybrid warfare' (the combination of conventional and unconventional means of warfare, as in the Ukraine since 2014) and also in the way that many wars now take place in civilian areas, leading to very high numbers of civilian as opposed to 'military' casualties in towns and cities (International Institute for Strategic Studies 2015b). Second, the sheer *number* of intra-state conflicts and their longevity is a new phenomenon when compared with the Cold War period, with one estimate being that more than one-third of countries have experienced an intra-state conflict, with two-thirds of these suffering from such conflicts over seven years or more (Gawerc 2006: 436). In 2014 there were 41 active conflicts (happily down nine on 2013 (International Institute for Strategic Studies 2015a)).

In addition, the overwhelming evidence that these 'civil' conflicts have had a habit of becoming regional problems, can be seen in the contagion effect in many parts of the world (Long and Brecke 2003: 5). This was evident in West Africa, where Liberia, Sierra Leone, and arguably Côte d'Ivoire were sucked into a regional conflict in the 1990s; Central and East Africa (where Rwanda, Burundi, the Democratic Republic of Congo, Uganda, Zimbabwe – among others – became, and still are, part of an arc of conflict after the Rwandan Genocide of 1994); in the Horn of Africa, with Eritrea, Ethiopia, Somalia, Sudan and Chad forming varying geometries of inter- and intra-state conflict since the end of the Cold War, and in many other parts of the world, as with the archipelago of Indonesia, where Timor-Leste is an integral part of a network of conflicts. The most obvious evidence of regional contagion saw wars in Afghanistan and Iraq between 2001 and 2009 break out more generally over the Middle East in Libya, Egypt and Syria (all 2011) and Iraq (which redescended into civil war with the withdrawal of American troops in the same year), in particular. At the time of writing only Egypt had resumed an uneasy calm under a newly instituted dictatorship.

The above, very selective, list of civil conflicts that have erupted since about 1990 gives a clue as to why war has become both much less devastating on one level while simultaneously being seen as more vicious. Arguably, contemporary civil wars are not nearly as terrible as the Indonesian civil war of 1965 (over 500,000 deaths), the Biafran war of 1967–70 (over a million deaths) and Cambodia 1975–9 (up to three million deaths). This is not to mention the appalling displacements and death tolls of the World Wars and the Chinese, Russian or Spanish revolutions (Kalyvas 2001: 110; Bellamy 2012). The worst casualty rates since 2011 have been in Syria, where well over 200,000 people have died. In part, better media coverage has made those wars seem more terrible than wars in the past, as have the huge refugee flows to Europe and elsewhere (one and half million to Lebanon alone).

One explanation of this seeming contradiction is that the fighting in previous decades was between industrial powers fighting traditional wars on an industrial scale, a 'settlement' of conflict through war, though even here the evidence is of civil wars always having been more numerous than inter-state wars (Mack 2006: 6–8). The answer maybe lies in changed perceptions of war itself. During the last twenty years or so of the Cold War, the Powers (and especially the United States and the Soviet Union) slowly realized that the use of force had its limits as a way of creating peaceful outcomes to conflict. At the end of it some even predicted that war would become 'obsolete', in an ironic echo of those who made the same prediction before 1914 (Angell 1910; Kaysen 1990). Maybe Mandelbaum was more accurate in saying that *major inter-continental* war may now be obsolete (Mandelbaum 1998), though again this is a dangerous prediction given historical precedent. What is clear is that in the early post-Cold War era, with the exception of the intervention in Iraq in 1990–91, the levers of 'hard' power were less willingly used by democratically cautious Western states that saw no evident national interest at play for them in many 'obscure' conflicts. Powers then aimed to try and bring about internally 'sustainable' ends to wars, ones that did not need continual revisiting. It became more fashionable, and responded well to a post-Cold War desire for a 'peace dividend', to discuss the use of 'soft' power (Nye 2005), and in particular a rise in the idea that conflicts could be talked down in workshops and off the battlefield. Joseph Nye defined this as follows:

> [Soft power] is the ability to get what you want through attraction rather than coercion or payments. It arises from the attractiveness of a country's culture, political ideals, and policies. When our policies are seen as legitimate in the eyes of others, our soft power is enhanced.
>
> (Nye 2005: Preface)

The peaceful transitions in South Africa, Mozambique, most of Latin America, even Israel/Palestine (with the Oslo Accords of 1993), and, most of all, in Eastern Europe, in the 1990s led to a swathe of books touting the benefits of 'Track One' and 'Track Two' conflict resolution, mediation and the like (see below for more discussion on these distinctions). There was, in the words of Samuel Huntington, a 'third wave of democratization' (Huntington 1993), though Huntington soon reversed his own optimism by declaring the 'Clash of Civilizations' (Huntington 1996). A real change began to happen at the end of the 1990s, as a result of the terrible events in the Former Yugoslavia and Rwanda in 1994–5. In the late 1990s, and even more so after that, there was a resurgence in the use of hard power in 'pre-emptive' mode in Afghanistan and Iraq and a refusal to negotiate with 'insurgents' in the 'war on terror'.

The truth of it is that the 'logic of violence' had not changed all that much, even if the response to it had. As a number of commentators have noted, violence has an internal logic of its own; it is 'rational' in the sense that it is often a response to local or more widely felt senses of injustice against governments, even resentment about global inequality or perceived cultural humiliation (Kalyvas 2006; Balcells 2010 and 2011). The 'Boko Haram' insurgency by Islamist militants in Northern Nigeria, Cameroon and Chad that specializes in massacres and kidnappings (most infamously of 274 schoolgirls in Chibok in April 2014) garners huge publicity and provokes terror well beyond the borders of its area of activity (International Crisis Group Africa 2014). The organization known as 'Islamic State' (ISIS or *Daesch*) uses both social and conventional media to publicize its aims through slick videos of beheadings and other executions as well as the encouragement of such acts in public places in Western cities like London, Madrid, New York and Paris. These 'spectaculars' also give it a platform in the Western media which has aided its recruitment of disaffected Muslim youth in the West, many of whom have died in Syria and Iraq.

As a result of the rising tide of violence since 2010, Western public opinion has played an important role in reducing the desire and the appetite for intervention by Western and other Powers, in spite of the often violent aftermath of the 'Arab Spring'. The civil wars (or 'revolutions') that erupted in, especially but not exclusively, Syria (2011–present) and Libya (2011), and the civil unrest and violent changes of government in Egypt and some Gulf states, added to already unstable and violent situations in Afghanistan and Iraq. The aftermath of American and other troops being withdrawn from Afghanistan and Iraq after the election of President Barack Obama in 2009 and a Conservative government in Britain in 2010, led to a reduction of the interventions by American and British administrations wedded to 'liberal internationalism' or 'neo-conservatism'. But it did not noticeably lead to a reduction of violent conflict or an improvement of levels of economic, social and political development in those countries as 'insurgent' forces took advantage of the power vacuums thus created.

Some writers believed that the UN and some major Western states had embarked after 2001 on a policy of 'international pacification' and 'pre-emptive regime change' (Duffield 2007 is one such example). It might be more accurate to say that a belief in the use of hard power, as opposed to softer versions, as a way of bringing about an end to conflicts in developing (and more 'developed') countries fluctuates in effect in line with optimism or pessimism about the prospects for different tools to end conflicts and the self-confidence of the underlying belief in prospects for a 'liberal peace'.

In trying to fashion responses to conflict both in the 1990s and since, much ink has been spilled trying to show the differences between various possible approaches. Conflict 'settlement', 'management' and 'resolution' have often been referred to as the 'three generations' trying to bring about peace. However, it must be understood that different approaches are often tried simultaneously or in sequence depending on the conflict being dealt with. So any attempt to be prescriptive about the use of such approaches has to be cautious and hedged around with caveats. History has an unfortunate habit of proving wrong any linear explanation of 'what is happening'.

So why attempt to define such terms? The reason is because they are so described in much of the literature on conflict and because they tend to coincide with particular approaches to international politics more generally. But first we need to put these concepts within the broader emerging paradigm of 'peacebuilding'.

Conflict 'settlement', 'management' and 'resolution'

There has been a long-standing tradition in international relations and conflict analysis of saying that different practical approaches to dealing with conflict can be seen as existing in parallel with certain 'paradigms' of international relations theory. This was particularly striking in the 1980s and early 1990s in the so-called 'inter-paradigm' debate in international relations. This rather simplistic thinking has been superseded by a much more subtle conversation about 'critical', even 'post-modern' theorizing about the nature of international activity. These latter thinkers have in particular attacked the ethos of thinking that 'we' can intervene to resolve 'their' conflicts, a way of thinking that has accelerated with the recent actions in Afghanistan, Iraq and the Middle East more generally. We will touch on these issues below. They are nonetheless key ideas, for as Jeong points out, all of these techniques and theories aim for a 'just outcome that is not only fair to participants but also meets the broader ethical concerns of a given society, and those of humanity' (Jeong 2008: 243). That is a tall order and unsurprisingly there are many who scoff at its practicality.

However, the terms that were developed then are still currently and widely used, so we need to understand how they might apply to particular ways of dealing with conflict by theorists and practitioners alike. It might also be mentioned that this is not a *linear* debate. Conflict analysis has its fashions and obsessions like every other field of human endeavour, and certain theorists and practitioners use the terms we will now explore in ways that defy exact categorization. There is no clearly and widely accepted *one* approach that all in this field agree on, just an ongoing, circular debate (Ramsbotham et al. 2011 give a good overview of this). Indeed we now have collections that point to myriad definitions about what is meant by even the expression conflict 'resolution' (Sandole et al. 2009; Allen Nan et al. 2012; Wallensteen 2012).

We can break this debate down in many ways, but the authors think it most useful to look at it first in the simple terms of 'top–down' or 'bottom–up', and then look at how that might translate into more precise terminology.

Top–down/bottom–up?

By *'top–down'* is meant all of those strategies that have been deployed by third party representatives of a state or IGO. This is often referred to as 'Track One' diplomacy. These kinds of action are in most cases undertaken before the conflict/war has come to the stage of a ceasefire. Sometimes this can be supplemented by 'Track Two' action, which is usually conducted by unofficial players, such as NGOs or officials acting in an unofficial capacity and behind the scenes, or even academics acting entirely on their own without any form of support.

Track One has been given huge prominence by the activities of American Presidents and their advisers (like Henry Kissinger) in the Arab–Israeli context since the 1970s, starting with the Camp David discussions of 1977 (Touval 1982). The main advantages of Track One are the legitimacy such discussions provide; the resources that can be marshalled in support of any agreement (military and economic aid, for example, as has been the case of Israel and Egypt since the late 1970s), and the guarantee of media interest in, and usually support for, such efforts. The disadvantages are that the Powers will use such occasions for their own interests and will cajole, bully or otherwise intimidate the parties and not in any way transform their basic unhappy relationship.

By 'bottom–up' is meant the introduction of a methodology for the local participants and victims of war or conflict to resolve their own problems using their own methodologies and systems. In many cases, but not all, these have indeed been elaborated locally with minimal outside help. Such actions usually occur after the fighting has ended, though hostilities may still simmer and flare up. This approach was particularly popularized by practitioner-theorists like John Burton and John Paul Lederach (Burton 1979, 1997; Lederach 1997). Outsiders in this way of framing conflict help to 'facilitate' the reduction of violent conflict by the locals themselves, and do not impose outside ideas or practices on them. Both the thinkers above assert that conflict is caused by frustrated 'human needs', not by any innately violent tendencies. Before trying to deal with it, we need to 'capture the complexity of conflict' (Sandole 1999).

Conflict settlement

'Settlement' is usually seen within a power-based realist paradigm where force is the arbiter. This can include 'mediation with muscle' as in the context of the United States using its power to broker agreements over the Middle East (as with Kissinger's mediation at Camp David, 1977, cf. Touval 1982).

Aim

Conflict settlement's aim is order, and often does not shrink from a 'realist' use of military force, sometimes on a huge scale, as in the First and Second World Wars, or from zero-sum outcomes. Realists take the view that all peace is a lull between wars and that trust between states and nations is a plausible but unlikely scenario (Mearsheimer 2001; Waltz 2001). They also deny that there is any such thing as an 'international society' with clear rules and processes, but that if such a thing exists it is due to a temporary balance of power and the dominance of a hegemon for a given historical period. Indeed, it has been pointed out that 'War appears to be as old as mankind, but peace is a modern invention' (Sir Henry Maine, quoted by Howard 2000: 1).

Problems

Conflict settlement often does not work, in the sense that only the maintenance of force keeps the 'lid' on the problem but does not address the underlying problems, indeed cannot, as states are always getting ready for the next war. We might also say that hard power is difficult to use when the areas where it has to be applied are ones in a state of civil war ('realism' assumes a functioning state system on the whole) and the conflict has a regional nature. As the United States has discovered in Afghanistan (2001–present) and Iraq (2003–present), the most formidable military machine can get very bogged down in an area wracked by an insurgency with no identifiable 'head'. As Iraq also shows, the idea of 'ending' a war by force can often create a bit-ter illusion. Wars that are finally 'ended' by force alone are in a dis-tinct minority and arguably always have been (Coker 1997). Even the hardest of realists understands that, as Clausewitz et al. (2012) com-mented, in war all is *'friktion'*, which might be translated as 'it's easy to get into, but unpredictable in the extreme once engaged upon' (see, for example, Gray 2005). The ethical issues for such thinkers and practitioners are clear, and they are largely ignored. So the main trou-ble with this approach has been highlighted by the recent wars in Afghanistan and Iraq. The 'locals' do not particularly like being forced to make peace, or to desist from war. One true believer in the use of force, Donald Rumsfeld, US Deputy Secretary of Defense under President G. W. Bush, admitted in November 2001 that:

> [O]ne of the lesson's of Afghanistan's history, which we have tried to apply in this campaign, is if you are a foreigner, try not to go in. If you go in, don't stay too long, because they don't tend to like any for-eigners who stay too long.
>
> (Roberts 2009: 29)

Conflict management

A slightly less, but still essentially realist, view is taken by conflict 'managers' who believe that it is usually impossible to do anything but try and grease the wheels of a conflict to make it less violent through

Plate 9 *A Tamil Tiger cemetery in Sri Lanka: third parties have struggled to gain the trust of both sides in mediation efforts.*

the judicious use of a combination of hard power, soft power (usually economic sticks and carrots) and peacekeeping forces. Roland Paris argues that such 'management' approaches can encompass 'liberalization', which he explicitly equates with 'democratization', a dual concept which we have linked to the 'liberal peace' theory and will be explored further in the next chapter (Paris 2004: 5). The ethical problem with this is of course that the 'targets' may not want to be so dealt with.

A feature of such approaches is also the use of 'Track One' mediators, usually in the form of 'good offices' or 'mediation' between leaders of insurgent groups, the established 'state' and any other regional players, a time-honoured use of a third party to help oil the wheels of a negotiation between states. Scholars have calculated that there have been visible mediation efforts in over 70 per cent of conflicts since the Second World War, and that does not take into account more discreet efforts (Bercovitch and Fretter 2004). Good recent examples in the 2000s include mediation by African Heads of State in the Democratic Republic of Congo and in Liberia. In the post-Cold War period mediation has come to also mean 'an extension and continuation of peaceful conflict management' (for a robust defence of this see Bercovitch, in Zartman and Rasmussen 1997: 127 and *passim* and Bercovitch and Rubin 1992). It should be noted that the expression 'mediation' can also be used without 'muscle' in the conventional sense of the latent or actual use of power. Religious leaders often have huge influence that they wield in a public or private capacity (Little 2007; Tenenbaum 2011). The degree to which this influence can be brought to bear can depend on imponderables like personality and credibility, the role of Archbishop Desmond Tutu in South Africa being one notable example (Sandal 2011).

Aim

Minimization, but not necessarily elimination of, violence; stability insofar as it can be achieved, and 'realist' expectations of the difficulty of getting people to agree on underlying causes, symptoms and solutions to their problems. The recent resurgence in the United States in the study of conflict management has much to do with the feeling that the optimism of the 1990s was misplaced and 'recognises that the global environs are both conflict-ridden and often bloody, and it harbors no illusions that murderous enmity will suddenly give way to universal comity' (Solomon in Crocker et al. 2007: xi).

Problem

Is conflict management too centred on domestic US obsessions (the 'War on Terror', etc.), so we might question as to whether it therefore sings to a tune not appreciated by others? It does nonetheless ask some very important questions about whether democracy is the 'answer' or stability the main aim and also proposes interesting ideas about new forms of sovereignty and other ways of organizing societies to make them less war prone (Ottaway and Marten in Crocker et al. 2007: Part VI).

Conflict resolution

Often delivered by 'Track Two' intermediaries, who can be academic groups, NGOs or local community activists, conflict 'resolution' is usually held to mean an attempt to fully (or mainly) resolve the underlying root causes of a conflict, by hopefully resolving (or indeed transforming) the relationship of the parties so that they develop an

entirely new and peaceful relationship. The theoretical and quite a lot
of the practical impetus for this idea can be traced back to the early
'conflict' and 'peace' researchers and practitioners of the 1960s and
since. These included Elise and Kenneth Boulding and John Burton,
as well as the scholars who worked with him and used his ideas, like
A. J. R. Groom and Christopher Mitchell (Mitchell 1981, 2014). This
heterogeneous group developed 'problem-solving' workshop techniques that were, and indeed are, claimed as the best tools for examining 'intractable' conflicts, of which the Israel–Palestine conflict is
the best single example. Institutionally, and also one linked to this
particular conflict, one of the best examples was the 'Oslo Process' of
the early 1990s, which seemed at the time to be capable of addressing
the underlying problems between Israel and Palestine (Corbin 1994),
but in 2015 was seen as a false dawn (see Box 4.2).

Aim

Full airing of differences, usually in a closed and confidential environment, with no preconceived agenda and a belief that problems and
parties to conflict can be 'transformed' so that a future relationship
will be better, less conflictual and ultimately stable (for a wider summary see Miall et al. 1999: 58). It purports to see *all* parties to a conflict in an equal light and does not exclude any potential solution to a
conflict situation. So, for example, the activities of the Centre for Conflict Analysis (established by John Burton among others) did not seek
to impose a pre-ordained solution, but led the parties to better communication of their needs (Burton 1997 *passim*; Williams 2005). Ideally
it will also 'highlight the wider social and political sources of a conflict in seeking to break the perpetuating cycle of oppression and
resistance' (Jeong 2008: 244). One of the best summaries of the characteristics and methodologies used can be found on the Berghof Foundation's website, a resource handbook that is constantly being added
to and updated (Fischer and Schmelzle http://www.berghof-handbook.
net/). It is worth quoting their definition at some length:

> First, war as an instrument of politics and conflict management can and
> should be overcome. Second, violence can and should be avoided in
> structures and relationships at all levels of human interaction. Third, all
> constructive conflict work must address the root causes that fuel conflict.
> And fourth, all constructive conflict work must empower those who
> experience conflict to address its causes without recourse to violence. In

short, conflict transformation must provide those who experience violence with appropriate and innovative methods and approaches, and assistance by a third party if necessary. Ultimately, this is about changing individual attitudes and addressing the issue of structural reforms.

Problems

An often naïve belief in human perfectibility; a belief that 'basic human needs' can be identified and rectified (Mitchell 1981; Burton 1990). One of the problems with analysing this approach has been the secrecy that is inherent in these processes, that has led to few of the case studies being written up for public consumption. This is now being somewhat addressed by, for example, examining 'transfer effects' that look at how Track Two experiences can be of use in Track One discussions and vice versa (Fisher 1997, 2005). Another more intractable problem is that it is seen by some as has having been demonstrated to be just another, if more subtle, use of power by the West to coerce the 'rest'. The idea that a number of parties to a conflict can *ever* be given equal attention is also naïve, especially as the international system as now constituted gives a clear priority to *established* nation states, not peoples seeking secession or self-determination. It has also been suggested that the insurmountable barrier to successful conflict resolution lies in the impossibility of understanding another culture, which is an 'ontological' barrier to any real possibility of transformation, a difficult charge to refute as conflict resolution teams rarely even understand the language of those they are dealing with, in developing countries in particular (Vayrynen 2001; Avruch 2003; Jabri 2007; Ron 2009). Another criticism that could be directed against such 'naïvety' is that in some cases it could be argued that success in transformation is only possible because some form of 'settlement' was already going on. The successful transition in Ethiopia/Eritrea in 1993 (states that subsequently resumed their conflict, 1998–present) was maybe because the military victory of the Eritrean and Northern Ethiopian insurgencies had brought about a radically new situation.

All of these categories of third party intervention have as a central aim the improvement of communication, whether it be to signal war or threat, or to enhance understanding of the parties' views, needs and wishes. Much of the theory and practice in third party intervention has been influenced by what could be described as the Western, legal, rational perspective. It has also been influenced by theories of bargaining and rational choice – some of which have limited application in the real world. And, as several analysts (Keashley and Fisher 1990; Webb et al. 1996) have stressed, multiple conflict transformation approaches can be pursued together, or one at a time, or in different orders, depending on the conflict in question. This has also been described as a 'multi-track' approach (with 'Track Three' being local community groups), justified when, as is usually the case, the conflict is particularly complex. What we tend to see is a hybridity in peacemaking whereby a variety of initiatives – both internationally and locally inspired – are simultaneously deployed.

Hence the authors, and many others, believe that the more overarching category of 'peacebuilding', leading to conflict 'transformation', is ultimately a more useful description of the process for creating a lasting peace, above settlement, management or resolution. Conflict settlement and management can be likened to holding a lid on a boiling pot, not attempting to turn the heat down, and conflict resolution has its own problems, as itemized above. So 'peacebuilding' is also often referred to as a 'third generation' approach to dealing with conflict (Richmond 2006) as we grope towards a successor to both conflict 'settlement' and conflict 'management', again because the underlying attitudes, causes and structures of the conflict need to be properly addressed.

Hence, it is on the ideal approach to dealing with deep-rooted violent conflict that we will concentrate, that of 'transformation', as we believe that should be the central aim of all approaches to conflict, even if not to the exclusion of the others, which are still the most widely used. It is therefore our preferred term, but we are aware that no term can be satisfactory, given the extreme complexity of many conflicts in developing countries and the limited resources available for dealing with them (see Chapter 1 for more discussion of these distinctions). Mitchell also argues that 'transformation' would require that enemies become 'partners, colleagues or, finally, friends so that their relationship can be said to be genuinely "transformed"' (Mitchell 2014: 190). Diana Francis (Francis 2002) gives another interesting description of one approach to the practice of conflict transformation.

The issues faced: the need for 'mapping' and 'deconstructing' conflicts

So what kinds of conflicts can be potentially 'transformed', and are they amenable to the kinds of theory and practical tools that we have

at our disposal? It is now, for example, widely believed that we are
dealing with conflicts that are not between *unitary* societies (usually
states), as was believed to be largely the case between 1939 and until
about 1990, but now within ethnically or ideologically *divided* societies. A number of theorists have put this down to the resurgence of ethnic violence (Gurr and Harff 1994, 2000; Kaufman 2001). Certainly,
compared to the period before 1990, one of the areas most identified as
being truly 'new' in the post-Cold War conflict mix is the huge rise in
what is termed 'ethnic violence'. Ted Gurr was one of the first to recognize this phenomenon in the context of conflict studies, but of course
there was a huge discussion of nationalism and self-determination well
before this (for an excellent introduction to this and the links to the
ethnicity question see Oberschall 2007: Chapter 1).

We might nonetheless question this now practically embedded wisdom
about the differences between pre- and post-1990 (Kaldor 2012). First
there were many civil wars *before* 1990 (Kalyvas 2001; Newman
2004). And, very unfortunately, ethnicity has often been assumed since
1990 to be a key component of a conflict situation both for the participants and for outside supposed 'conflict resolvers'. This 'framing' can
have disastrous effects, when it arguably is not the determinant factor
by any means. Kaufman (Kaufman 2001) wrongly sees ethnicity as the
root of most recent conflict in Eastern Europe for example. A classic
example of this was in Bosnia (Campbell 1998a, 1998b), but such
misjudged preconceptions have been widely and unfortunately repeated.
It is far more likely that economic or ideological factors underpin much
'ethnic' conflict (Nordstrom 2004; Pugh and Cooper 2004).

The problem is arguably worse in developing countries where the
obsession with 'identity' has tended to obscure the continuities of
local (for example African) history and traditions that pre-date the
colonial era. Our tendency to forget about historical precedent and
memory has led us, in Richard Werbner's words, to 'mark the end of
an epoch falsely by placing a break where none exists' (Werbner in
Werbner and Ranger 1996: 5). The result has been to both marginalize developing countries as mere adjuncts of Western historical
experience, 'post-colonial' child-like entities without local agency
capable of creating their own destinies. But, most significantly, the
creation of states themselves has *always* been a violent process.
Massacres, civil wars and the like have attended the birth of all the
'classic' states of the past (Tilly 1985; de Tocqueville 1856). Edward
Newman and others ask whether the big difference now is that the

'statebuilding' project (see Chapter 5, pp. 172–4) is now mainly different from that of the past in that these processes are supposedly being shepherded by IGOs like the UN with the aim of liberal state-building (Newman 2013: 142–3).

So, although the analysis of all conflict as 'ethnic', or even 'civil', is a neat and workable shorthand in many conflicts, it is not sufficient (Brubaker and Laitin 1998). As Oberschall puts it, '[t]hese theories lack specificity and context'. His preference is for what he terms 'conflict and conciliation dynamics', which essentially seek to 'map' the conflict by looking at issues, players and their strategies, and implementation (Oberschall 2007: 29). This idea of mapping a conflict has also been used quite widely in conflict resolution and transformation approaches (including those of Mitchell and Webb 1988; Burton 1990; Fisher 1997; Kelman 2005, etc.). For all these theorists we might suggest a widespread belief that a first step *has* to be to 'deconstruct' a conflict before it can be properly understood and analysed, as a precursor to identifying possible solutions to it. (See also Chapter 1 for more discussion of this vital theme.)

As we stressed in the Introduction, an important input into this school of thinking has been the work of Edward Azar; who believed that 'protracted social conflict' (Azar 1990) needed 'problem solving', not military solutions. Azar's approach was one that accepted much of what has been written by theorists of 'basic human needs' like John Burton. Their aim was to get a wider acceptance that most conflicts were due to 'the prolonged and often violent struggle by communal groups for such basic needs as security, recognition and acceptance, fair access to political institutions and economic participation'.

If this is true, then unravelling the motives for perpetrators, victims and interveners alike is far more daunting than dealing with mere ethnic violence, bad as that obviously is. It is obvious that the use of violence has multiple motivations, with correspondingly difficult decisions to be made about how to 'map' these motivations. A key, and growing, element of war has been the reported use of sexual violence, with increasing debates about motivations and potential solutions to the use of rape in war (see Enloe 1993; for a recent, if controversial, summary, see Thornhill and Palmer 2000). We might, however, also point out that sexual violence in war was as much a part of the repertoire before 1990 as after it. Historians such as Anthony Beevor have catalogued the widespread use of rape by Russian soldiers in Berlin in 1945 (Beevor 2002) and Edward

Newman has also made a similar point, again in criticizing the 'new war' thesis. (Newman 2004).

Azar's approach allowed for the analysis to be on a variety of levels – 'identity' at the 'communal level' with an analysis of the 'deprivation of human needs'; the crucial factor of the 'governance and the state's role' and the need to look at 'international linkages' (Azar after Ramsbotham et al. 2005: 84–8). All of these need a long-term process that is often beyond the means of 'Track Two' teams and beyond the mandate of 'Track One' mediators. We cannot yet be said to be anywhere near resolving this conundrum (but for some understanding of it see Fisher 2005).

Strategies of conflict resolution and transformation

So although purists will immediately see flaws in such a narrative, we can therefore, cautiously, describe most attempts by (usually but not always) outside parties at mitigating or 'solving' conflicts through two major series of approaches: 'top–down' (conflict settlement and management) and 'bottom–up' (conflict resolution or transformation). We can also suggest there is now in place an overall idea of 'peacebuilding', defined by Miall, Ramsbotham and Woodhouse as 'the attempt to overcome the structural, relational and cultural contradictions which lie at the root of conflict in order to underpin the processes of peacemaking and peacekeeping' (Miall et al. 1999: 56–7). This can be seen in terms of actions by states or IGOs, and as such is it usually used in the 'reconstruction' attempts we will describe in the next chapter. But the emphasis in the rest of this chapter is on 'peacebuilding from below', which is what the term often means in the context of conflict resolution, transformation, or attempts at reconciliation, all of which often, though not exclusively, are used by changing the attitudes and practices of the base.

To put that into simple language: how would you, our reader, think it best to deal with: a) two warring factions within a government, or b) the reintegration of people who had been killing your neighbours or family? Would you advocate a policy of justice (or revenge) for paramilitary forces that may have destroyed your village or killed your relatives, or their 'reintegration' into society (as suggested in the next chapter), or getting them to undergo a process of much more personal reconciliation?

Reconciliation: memory, truth-telling and forgiveness

One way to bring about a lasting peace is now seen as being through 'reconciliation', either through the mechanism of 'truth commissions' (TRCs), often referred to as 'restorative' or 'transitional justice' (probably the more used of these terms nowadays), or through the use of a much older institution, the 'War Crimes Tribunal' (WCT), often referred to as a form of 'retributive justice'. The first usually assumes a non-retributional and non-jural outcome; the latter assumes the use of some form of sanction, including imprisonment, or even the death sentence, for 'guilty' parties (Rigby 2001; Bell et al. 2007). We will discuss them in turn, but first look at what they can be said to have in common in addressing conflicts in developing countries.

Both of these institutional frameworks attempt to work based on the proposition that we need to establish the 'facts' of any conflict, including the hardest of all: those to do with killing, torture and other massive abuses of human rights. This, it is assumed, will somehow help both victims and perpetrators to come to terms with their experiences and actions and help them to understand, possibly forgive, and hopefully 'move on', to a better and more functional relationship. It also assumes that the bringing of 'justice' will indeed reduce the levels of violence and contribute to peacebuilding (Lambourne 2009). These propositions are drawn from essentially three sources:

- First, from psychological insights that a 'talking cure' (Freud's term) will enable individuals and societies to make a transition from hostility to peace through the establishment of clearly identified memories and contending 'truths' and that this will then lead to a better ability to deal with collective and individual trauma (Jeong 2005: 155; Fierke, in Bell 2006; Fierke 2013)
- Second, it bears more than a passing resemblance to ideas in conflict resolution/transformation theory and practice that a 'ventilation' of grievances and an establishment of the history(ies) of a conflict will help establish a new point of departure for the players and thus a transformation of their relationship discussed above, and
- Third, its proponents often link it explicitly to 'liberal peace' arguments that 'transitions to democracy' necessitate an adoption of market reforms, constitution building and the establishment of

a rule of law and a consideration of 'what to do about the past'. Justice and accountability will go hand in hand with prospects for a stable peace (Hayner 2002: xi, Chapter 7).

The first attempts at setting up TRCs were in Latin America (Chile, El Salvador and Argentina in particular), to attempt to draw a line under the 'dirty wars' of the Cold War era (Whittaker 1999: Chapter 2 on El Salvador; Hayner 2002). After 1990 the most prominent examples have been in South Africa after the end of Apartheid (see Box 4.3 for more detail), the Rwandan Genocide of 1994 (Prunier 1995; Corey and Joireman 2004; Moghalu 2005, also see Box 4.4) and in Sierra Leone after the Lomé Peace Accords of 1999 (http://www.trcsierraleone.org/drwebsite/publish/index.shtml). Suggestions have been made for similar commissions in the Former Yugoslavia, in Northern Ireland and elsewhere.

Scholars who have assessed the successes and failures of TRCs have come to mixed conclusions (Olsen et al. 2010; Gready 2011). Nonetheless, the aftermath of the civil wars in Sri Lanka, Myanmar and other parts of the Asia–Pacific region has led to a renewed interest in mechanisms of transitional justice, especially given the inter-communal nature of the violence and the often brutal attempts to 'manage' or 'settle' them (on Sri Lanka see: Sriram in Jeffrey and Kim 2014; on Myanmar see International Commission on Transitional Justice http://www.ictj.org/our-work/regions-and-countries/burmamyanmar). Transitional justice is now a standard part of the tool kit of peacebuilding.

One way of assessing the usefulness of such attempts is to analyse the underlying essential elements of such thinking, and especially the ideas of 'memory' and 'truth'.

Memory, history and 'truth'

An essential element of reconciliation lies in having common, or at least mutually comprehended, memories of what happened in a conflict or war that needs to be reconciled. History is a powerful weapon in the hands of those who would establish their version of the 'truth', but also a potentially even more useful tool for shedding light on the wrongs experienced by all sides in a conflict, thus allowing them to (potentially at least) 'move on' in their relationship. The lack of a historical dimension to much of the study and practice of international relations can be identified as a grave flaw in policymaking and actions in a large number of recent and older conflicts and wars (Williams et al. 2012). The role of historical factors in decisionmaking by the United States government about how to 'reconstruct' Iraq after 2003 will be dealt with in Chapter 5 (pp. 169–72 and 187)

The aim of much conflict transformation and reconciliation is thus to establish both the 'individual truth' of what happened, but also to interrogate what might be called 'collective' truths. This, it is hoped, will then lead to mutually acceptable changes in the versions of the 'truth' that have long dogged the peoples involved in a long-running feud. If, for example, the Irish and the English or the English and the Scottish people persist in 'remembering' their past histories as exclusively ones of mutual persecution, with 'signposts' along the way that encourage a prolonged feeling of mutual grievance, or even hatred, there is little chance of them being able to make lasting peace with one another. It might be reasonably asserted that the popular film *Braveheart*, which portrayed the heroic and allegedly largely blameless William Wallace as fighting a guerrilla war against the heartless and brutal English monarchy, did little to foster Anglo–Scottish understanding, and even contributed to the much remarked upon resurgence of Scottish nationalism since the 1960s (Webb 1978). Equally, folk memories of Cromwell's massacres at Wexford and Drogheda in the 1650s or the Battle of the Boyne, won by William of Orange against the rump of James III's army in 1690, have been constantly evoked in nationalist propaganda by Republicans and Loyalists alike in Ireland to justify horrific atrocities. Even more recently the memories of the Great War of 1914–18 have been used in Ireland to justify Protestant violence against Catholics in the North (Mac Ginty and Williams 2007).

Lest this be thought too British–Irish a set of examples, it should be noted that in the Former Yugoslavia the Croat and Serb leaderships both evoked historical grievances dating back to the fourteenth century in their pursuit of ethnic cleansing and other brutality. History is indeed a dragon that it is dangerous to awaken! But if that is the case, can history be used as a way of reconciling and resolving conflicts, as we have suggested about Ireland?

Box 4.1

Transformation by historians

In 1914 Ireland had the characteristics of a developing state (mainly agricultural, very traditional cultural practices – especially through the dominance of the Catholic Church and a semi-colonial relationship to a 'mother country', the UK). Ireland is now a modern, even post-modern state with a typically capitalist economy and a liberal democratic political system, tied to a supra-national liberal capitalist regulatory system, the EU, a model often seen as 'perfect' by aspiring LDCs (least developed countries). Ireland experienced major political violence and civil war (especially from 1916–22) as well as a long-standing conflict with a UK-linked 'unionist' population in the North of Ireland. The dominant historical story told about Irish history was that of the victors of 1916–22, that 'loyalists' were political stooges of the UK (or even the 'English'), and that very few 'true' Irishmen had fought for the UK in the First and Second World Wars of 1914–18 and 1939–45. This has now been corrected politically by the 'Good Friday' Agreement of 1998 and other agreements, and an acceptance that 'unionism' is a legitimate political position. It has been rectified historically by a new acceptance that Irish soldiers from the Republican and Unionist traditions fought in the Great War, often together (Ferriter 2015; Walsh 2015), that there might well have been a 'constitutionalist' alternative to civil war and even exit from the British Empire (now the Commonwealth) by Ireland. Historians have played a large role in this (Bew 2009; Laffan 1999), acknowledged by Irish politicians like John Bruton (Taoiseach, 1994–7), Irish Presidents Mary Robinson (1990–97) and Mary McAleese (1997–2011), as well as Queen Elizabeth II and key politicians in the UK.

Jay Winter, in many ways the Anglophone father of memory studies, has written about the 'memory boom' in contemporary historical studies (Winter 2000). There has been a widespread rememorization of war in Western Europe as part of a process of reconciling the peoples of Western Europe who only a generation back (and for several before that) were regularly engaged in internecine warfare of the most brutal and uncompromising kind (Winter 1995; Mac Ginty and Williams 2007). The First and Second World Wars were not African or Asian affairs after all; they were purely European innovations, even if they had profound effects on the rest of the world. Many of the insurgents/freedom fighters in wars of decolonization (Algeria, Kenya, Vietnam, arguably Israel, are good examples) learnt how to fight their colonial masters by fighting alongside them in 1914–45. Their resentment at the pronouncement of adherence to 'democracy', and the evident hypocrisy inherent in

Plate 10 *French First World War cemetery: issues of commemoration can have long-lasting sensitivity.*

such claims, fuelled many uprisings (see for example Lewis 2007 on Kenya).

A number of questions arise immediately:

- How can, or should, these historical ghosts be put back to sleep? Is full/partial disclosure, or amnesia better? (Rigby 2001)
- Does the existence of both national mythologies and, in some cases, extremely antagonizing personal and collective memories matter in potentially fostering future violence?
- What if there are 'contested memories' (no agreement on what happened)? (Beneduce in Pouligny et al. 2007)
- Who has 'ownership' of a historical narrative and could therefore be trusted to write an account of it that would not (inevitably perhaps) be merely a version to suit the 'victors', or at least a dominant view of events, or put their spin on commemorative activities? (Mac Ginty and Williams 2007)

The case of Israel/Palestine is most instructive in this regard:

Box 4.2

History, Israel and Palestine

The Israeli–Palestinian conflict is steeped in historical controversy. The British and French Governments of the First World War have been roundly criticized for their roles in promising different groups and peoples the same areas of the Middle East in return for their help in defeating the Ottoman Empire. Hence 'Zionist' politicians were promised a homeland in Palestine in 1917 (the 'Balfour Declaration'), while simultaneously British officer T. E. Lawrence ('of Arabia') made promises to Sharif Hussein of Mecca that he could have Damascus and Jerusalem as the capitals of a new Arab state. Meanwhile the French and British (in the the Sykes–Picot Agreement of 1915 and the Treaties of Versailles, 1919, Sevres, 1920 and Lausanne, 1923) divided up what became Iraq, Jordan, Palestine, Syria and Lebanon between them. This has led to endless arguments about 'who owns what and by what right' in the Middle East and is a foundational problem in finding a solution. It was compounded by the circumstances of the foundation of Israel in 1947, after which large numbers of native Palestinians were expelled to neighbouring countries where they and their offspring still live and form the core of anti-Israeli feeling in the 'West Bank' and Gaza, with the eruption of frequent 'Intifadas' (uprisings) against Israeli dominance. Israel has had to submit to being attacked by multiple Arab armies on several occasions (1948, 1973 and in numerous cross-border incursions) and been forced (in its view) to carry out pre-emptive wars (1956, 1967, plus major incursions into Lebanon in 1982, 1994 and 2006). The whole region is thus affected. In addition to that, as US President Carter put it:

> People know if they are from a war torn country how difficult it is to sit down across the table in the same room with an adversary. Just consider about the Israelis negotiating with the Palestinians. It is likely adversaries will say: 'we cannot negotiate because we despise the other side too much. They have killed our children, they have raped our women, they have devastated our villages.'

Past and present grievances become as real as each other and therefore a solution needs to address such hurts. The Oslo Process of the early 1990s tried to do so by addressing the triple issues of the 'right of return', the status of Jerusalem and the borders of any future Palestinian state. Those are still the key issues today. The history and the present issues will need a global solution.

Sources: Ulrichsen (2014), Barr (2012), Fromkin (2000), Corbin (1994), Jones (1999).

The main issue is the existence of two opposing historical narratives. If we accept the Israeli narrative, which essentially sees Jews having fled the Holocaust in Europe and settling an 'empty' land and developing it where before there had been desert and poverty, to the benefit not just of the Jewish population but all those within and indeed without Israel's borders, then a picture of a blameless Israel being attacked by fanatical hoodlums makes absolute sense. If we accept the Palestinian mainstream view that they were expelled in large numbers from their homeland by an alien invader backed by the Great Powers in 1947–8, Powers which were themselves motivated more by their guilt about the Holocaust than any genuine sympathy for either Jews or Arabs, then obviously we can sympathize with the Palestinians. As historians have pointed out (notably Edward Said, see Ruthven 2003; Bunzl 2008), these 'mirror-images' or incompatible mythologies are the basis of many protracted conflicts.

'Ruptured histories'

The idea that history is 'dead' or that we have an 'end' (Fukuyama 1992) to it could not be less true than it is today. History informs most, if not all, of the enduring deep-rooted violent conflicts that afflict the developing world. Some of them, like those that pit Japan as a former imperial power against South Korea and China, are vital for understanding the future stability of the whole of East Asia (Miyoshi Jager and Mitter 2007; Kim in Jeffery and Kim 2014).

So what can be done? Should those identified as having responsibility for crimes of war be brought to book and put before a court of law, national or international, as mentioned earlier and often referred to as 'retributive justice'? Or should the miscreants be seen as victims as much as perpetrators and allowed or encouraged to reinsert themselves into societies that they once sought to destroy? We will now go through the arguments for both of these courses with the use of historical examples. It should be noted that there are significant differences that exist in the theory and practice of how a conflict can be de-escalated and a peacebuilding process given a chance to begin.

Some general issues need to be addressed first.

A conflict might be seen on one level as an inability to trust. Ramsbotham et al. put it thus:

> [l]egitimacy, acceptance and trust ... are integral to the functioning of
> any reasonably stable socio-political system, invisible and often taken
> for granted when differences are being settled relatively peacefully,
> but palpably lacking when they are not.
>
> (Ramsbotham et al. 2011: 206)

They also point out that 'one of the main obstacles to social and psychological healing is the cumulative hurt' (Ramsbotham et al. 2011: 207). So a spiral of distrust develops in an uncanny echo of the 'security dilemma' in international politics (for an explanation of this see Booth and Wheeler 2007). Because you believe that the other side has more weapons than you (or is less trustworthy), you arm yourself more (and trust less). The classic example of a global conflict is the arms races of the Cold War; a more current and local one the ever increasing use of more and more sophisticated small arms in pastoralist conflicts in Africa (Mkutu 2008; Wepundi et al. 2012).

Truth and reconciliation commissions

Box 4.3

The South African Truth and Reconciliation Commission (TRC)

The end of Apartheid in South Africa came after an all-white referendum in 1992 and the freeing of Nelson Mandela by the National Party, led by F. W. de Klerk. De Klerk voluntarily handed over power to a new majority administration of the African National Congress (ANC). The ANC has followed a very 'liberal' path since then, holding a number of elections, encouraging foreign multinational companies to set up and develop in South Africa and avoiding all of the worst excesses of neighbouring Zimbabwe by encouraging a process of reconciliation. This has been concentrated in the Truth and Reconciliation Commission (TRC), set up in 1995, which has heard hundreds of testimonies from former perpetrators from the South African defence forces and many more victims and their relatives. The idea has been that suggested by Archbishop Desmond Tutu, that the truth will heal the trauma of the past and that it will help in the 'nation-building' of South Africa. Some have argued that the process, which has never really addressed the problems of reparation for the crimes committed (although there is provision for this in the TRC's statutes) lets off the perpetrators too lightly. The Commission finished its work in 2002. The Government is pursuing a policy of gradual and cautious reform of land, for example, which it is hoped will help the reconciliation process along.

Sources: Rotberg and D. Thompson (2000), Hayner (2002), Williams (2006): 191–4, Gready (2011).

Another hard case: Rwanda

The question we have asked above can be summed up as: can/do TRCs help in dealing with the aftermath of deep-rooted violent

---

Box 4.4

Rwanda

By many calculations 500–800,000 mainly 'Tutsi' men, women and children were mas-
sacred by their 'Hutu' neighbours in 1994, and many women gang-raped, resulting in a
huge subsequent dispersal of the population that committed these massacres as the new
government, which was also ethnically mainly Tutsi, came to power and demanded ret-
ribution. The international community turned a largely blind eye to the massacres, in
spite of anguished pleas to the UN by the UN military commander in the country, General Dallaire (Dallaire 2005; Melvern 2006). The French Government (through 'Operation Turquoise') has been accused of harbouring the 'génocidaires'. The massacres
were of an ethnic nature but might be argued to have been about other issues, as the
two groups were practically indistinguishable. They were certainly about elite competition and demonstrated the power of a centrally organized bureaucratic state in coordi-
nating such events, as Nazi Germany had previously shown. The post-massacre
'Gacaca' courts, set up in 2002, have attempted to get lesser (that is to say not leading)
perpetrators to confront the reality of what they have done, in exchange for a lesser or
even no punishment. The post-genocide era has seen the imposition of another authoritarian state, and massive intervention by the Rwandan army in neighbouring Congo,
though with a lot less violence being committed within Rwanda. About 110,000 were
incarcerated awaiting trial in 2001. An International Criminal Tribunal for Rwanda has
also been set up by the UN but its effectiveness has been less than impressive in terms
of the numbers of those brought before it or punished, though important legal prece-
dents have been established, against sexual violence for example.

Sources: Straus in Pouligny et al. (2007), Kayumba and Kimonyo in de Zeeuw and Kumar (2006): 211–36, Booth – on
sexual violence in Rwanda – in Sands (2003): Chapter 5, Moghalu (2005).

---

conflicts? The South African case is a hard one in that unravelling
nearly a century of apartheid policies may well take generations. But
what about an even harder case, that of Rwanda?

What can we learn from such episodes for the future of reconciliation
attempts in developing countries? It is evident that in Rwanda, 'even
by strict definitions the episode qualifies as "genocide"' (Straus in
Pouligny et al. 2007: 123), so the aftermath of such events is that
much more difficult to deal with than, but also different from, the
long-drawn-out agony of South Africa. In both cases we could argue
that these were 'unforgivable' crimes. In South Africa the fight
against Apartheid came to be seen as the sole prerogative of the
African National Congress (ANC). Mahmood Mamdani has asked the
pertinent question as to whether 'if truth has replaced justice in South

Africa, has reconciliation turned into an embrace of evil?' He also points out that 'in Rwanda there are a lot of perpetrators and a few people who benefited; in South Africa there are a few perpetrators, but lots of beneficiaries' (Mamdani, quoted by Krog 1998: 146–7).

This conundrum has led many to ask if the reconciliation in either country is more than skin-deep. The post-Apartheid crime figures in South Africa (where hundreds of white farmers have been killed and the country has one of the highest murder rates of any country on the planet) suggest a society ill at ease with itself. So is reconciliation the 'lesser of two evils?' Mothers who lost their children have a different view of the idea that reconciliation can help us 'turn the page'. One such mother who testified to the TRC, a Mrs Kondile, commented that she could not forgive a South African policeman, Dirk Coetzee, for killing and burning her son: 'It is easy for Mandela and Tutu to forgive… they lead vindicated lives. In my life, nothing, not a single thing, has changed since my son was burnt by barbarians… nothing. Therefore I cannot forgive' (quoted in Krog 1998: 142). Others have claimed that, in the case of the TRC, even if the whole truth is not found out, a great deal is learnt and the process is therefore 'therapeutic' (Christie 2000: 173–5).

War crimes tribunals

Some have suggested that a better solution would be to opt for the forensic truth-seeking of a legal retributive tribunal, as was the case with the Nuremberg and Tokyo trials after the Second World War, the records of which are still taken by many as one quasi-'definitive' account of Nazi and Imperial Japanese war crimes (Sands 2003; Browning 2005; Friedlander 2008). There are many detailed discussions about the nature of both German and Japanese war crimes (on Japan see Rees 2012) and the subject has led to sharp exchanges between historians and political figures in Japan, with a notable contribution to truth-seeking from the Japanese historian Chihiro Hosoya (Hosoya et al. 2001). As we discovered in discussion with the excellent Japanese translator of the first edition of this book (Mac Ginty and Williams 2012), there are still deep divisions about how these issues should be interpreted. There is also now some scholarly discussion in English as to whether we can in fact compare the Nuremberg and Tokyo trial processes with any ease (Futamura 2008).

In recent times the International Criminal Tribunal for the Former Yugoslavia has drawn a huge amount of media attention and has led to the prosecution of many alleged perpetrators of war crimes in the Balkans, most famously of Former Yugoslav President Slobodan Milosevic and, in 2008, Radovan Karadzic, the Bosnian Serb leader (Williams 2006: 169–74; Donia 2014; Baker 2015). There is also a permanent International Criminal Court (ICC) that serves the international community as a whole, formally founded in 1998 and directed against individual misdemeanours, potentially also by peacekeepers (which is why a number of states have not ratified the treaty setting it up). Its website as of April 2015 tells us that 123 countries are full signatories and that:

> The ICC is a court of last resort. It will not act if a case is investigated or prosecuted by a national judicial system unless the national proceedings are not genuine, for example if formal proceedings were undertaken solely to shield a person from criminal responsibility. In addition, the ICC only tries those accused of the gravest crimes.
>
> (http://www.icc-cpi.int/, accessed 14 July 2015)

The ICC has been mainly used in the Congo, but most famously in the Darfur conflict in Sudan where the first ever sitting Head of State, Omar Bashir, has been accused of crimes against humanity.

If TRCs might therefore be seen as 'soft' option, as in the cases of Rwanda or Sudan for example, might 'retributive', legally enforceable, and punitive criminal tribunals not be a better option to help in the needed processes of healing described above? The use of the war crimes tribunal (WCT) has of course been explored in the context of developing countries and their conflicts. The International Criminal Tribunal for Rwanda (http://www.unictr.org/) has been in existence since 1994, just after the massacres, and has as its official brief 'to contribute to the process of national reconciliation in Rwanda and to the maintenance of peace in the region'. The criticisms of its breathtakingly slow pace and ineffectiveness have led to some commentators thinking it brings the rule of law into disrepute (Moghalu 2005). There is a similar one for Sierra Leone, the Special Court for Sierra Leone, which used the ICC's facilities in The Hague to prosecute Charles Taylor, widely accepted as being a prime mover in the horrors that afflicted much of West Africa (for more on this see Wierda 2006: 183–207).

One potential problem for such tribunals' effectiveness in developing country contexts is that their roots clearly lie in the Nuremberg and Tokyo Tribunals after the Second World War, when the circumstances of global politics were very different to now and when the justice being meted out was in the context of inter-state war, not in our predominantly 'new' war era. Another problem is that which can also be levelled against war crimes tribunals in the industrialized world – on occasion the needs of 'justice' and those of 'peace' are in distinct opposition. How can you expect a warlord to lay down his weapons if he knows he will be subsequently accused of war crimes and probably executed or imprisoned for life? Equally there is some sense in the US criticism of the ICC that it potentially criminalizes soldiers doing their best to do the bidding of the UN. Against this it could be argued that all 'proper' armies prosecute their troops for criminal behaviour, as have the Americans in Iraq and Vietnam for example, and that the above ICC stricture about it being a court of last resort answers that criticism. Maybe more seriously, the ICC was not used in the prosecution of those deemed to be war criminals in Iraq – the executions and other punishments inflicted there were by domestic judicial instances, and might be accused of being essentially Victors' Justice (Bass 2000; Williams 2006). We feel there is not enough evidence, for or against, to yet have a judgement on the effectiveness of such tribunals in the context of conflict in the developing world.

To summarize: the aims of TRCs and WCTs are to try and:

- Accommodate the vital role of historical memory. Each player in the conflict believes that history is on their 'side' and evokes it to justify ever increasing cycles of violence and revenge.
- To deflate this spiral the inverse process of historical truth-telling has to take place, whether though a formal system of justice ('retributive' justice – WCTs for example) or a non-judicial process ('reformative' justice, such as TRCs).

It is important to note that the main aim of both TRCs and WCTs is to make an *attempt* to address the above aims. Hayner quotes Michael Ignatieff's caveat that 'The past is an argument and the function of truth commissions, like the function of honest historians, is simply to purify the argument, to narrow the range of permissible lies' (Ignatieff in Hayner 2002: 25). This has given rise to frustrations and potential future conflicts.

Conclusions

Anthony Oberschall sums up the substance of the issues dealt with in this chapter in two sentences: 'Collective Myths are the enemy of truth and justice… conflict management and peace building are hampered by crisis framing, the victim syndrome and denial' (Oberschall 2007: 2). While we might disagree about the exact wording, this seems to us a fair summary of dealing with the kinds of problems faced in trying to manage or resolve most conflicts in developing countries, as it has also been in some in the 'West' like Northern Ireland, other parts of the European Union (the Basque Country most notably) as well as the Former Yugoslavia and the Former Soviet Union. So a first conclusion has to be that the problems of developing countries are not *sui generis*. They can happen anywhere, wherever there are people struggling to find economic, political or cultural identity and meaning. One of the worst obstacles that needs to be overcome if we are to improve what is a dismal track record of dealing with complex deep-rooted conflicts is to assume that they are easy to resolve, but also that we should not try.

Second, we might also ask whether there are lessons to be learned from looking at *non*-conflict situations, as any thesis has to be falsifiable. Of course this is the basis of liberal peace arguments. Such theorists would say that the effective deployment of democratic institutions is what makes a state succeed in avoiding destructive conflict and enhancing successful development (Paris 2004; Kaldor 2012). Although such considerations are very interesting, we feel we need in this book to concentrate on attempts to address what does *not* work. What seems clear to us is that the official discourse of participation and empowerment through democracy often falls far short of the reality. In most cases the evidence seems to be that such discourse and practice can actually depoliticize and demobilize the poor, who are the subjects and objects of much conflict behaviour (Moore 2001; Williams 2007).

So the short-term effects of such peacebuilding processes can be perverse. Many have accused the South African and Rwandan TRCs of asking the impossible – forgiveness by victims for unforgivable crimes. An area that needs much more research is in the long-term effects of such attempts and what might be done to improve the prospects for success. The phenomenon of 'inter-generational violence' has been noted. Although the evidence is often anecdotal, it would appear that, for example, in areas of the Former Yugoslavia where there were widespread atrocities during the Second World War, there were higher levels of 'revenge' violence in the 1990s (Dragojevic 2013). One area that is now being explored is that of the effects of conflict on (even) unborn children (Devakumar et al. 2014) and the possibility of education for succeeding generations. The German example of education changing wider attitudes is often cited (Dierkes in Cole 2007). Likewise the negative evidence of such educational attempts not being made in the teaching of history in Japan, Korea and China have also elicited widespread comment (Bleiker 2007; Miyoshi Jager and Mitter 2007; Yoshida 2007). Not many analyses have yet been done on developing country cases, however.

Third, we must ask whether an outside party *should* interfere in the warfare/conflict raging within a developing country, of which there are many examples. And will their efforts stand any chance of succeeding, and why or why not? As John Darby (2006) has put it: 'The signing of a multiparty peace accord is unlikely to end violence.' We need to look beyond such 'institutional' concepts as conflict settlement and management, discussed in this chapter, and democratization, disarmament and the reintegration of former combatants, reconstruction and institutional change, which we will discuss at length in Chapter 5, to see how deeper mechanisms can be used to try and coax whole populations and especially elites towards a more gentle and less confrontational way of solving their inevitable societal conflicts.

Peacebuilding, conflict resolution and the transformation of societies and individuals from the state of violent conflict or war is never going to be an easy task. It is difficult in developed countries like Ireland; it is all the more difficult in conditions of underdevelopment, poverty and state collapse. Chapters 5 and 6 will address these vital corollary issues. Can development 'from without' help societies become more economically and politically stable so that peace can have a chance of succeeding?

Summary

- Theories of how to deal with conflict (settlement, management and 'transformation') are often seen as dependent on the observer's underlying theory of international relations.
- Most, but not all, conflicts in developing countries are civil wars.

- Peacebuilding and conflict 'transformation' approaches can be 'top–down' or 'bottom–up'.
- Protracted social conflict cannot simply be explained by reference to 'ethnic' differences.
- Reconciliation within divided societies can be attempted by internally organized jural or non-jural tribunals, or by international war crimes tribunals.
- We need to understand the historical roots of conflicts in order to have any chance of resolving them.

Discussion questions

- Is it true to say that 'new wars' have now largely replaced the 'old wars' of the era before 1990?
- Should peacebuilding be approached in a 'top–down' or 'bottom–up' way?
- Can the International Criminal Court provide impartial justice or does it rather prosecute individuals from states that are vulnerable to pressure?
- Are commissions such as the South African Truth and Reconciliation Commission a better way to bring about peace than war crimes tribunals?
- What role do you think the 'history' of a place plays in helping us understand the causes of and potential solutions to conflict and war?

Further reading

(Possibly) the best introduction to the entire field of conflict resolution and management can be found in J. Bercovitch, V. Kremyenuk and I. W. Zartman (eds) (2008) *The Sage Handbook of Conflict Resolution*, Thousand Oaks, CA: Sage Publications; and, on conflict management, J. Bercovitch and J. Z. Rubin (1992) *Mediation in International Relations: Multiple approaches to conflict management*, London: Macmillan; or C. A. Crocker, F. O. Hampson and P. Aall (eds) (2007) *Leashing the Dogs of War: Conflict management in a divided world*, Washington, DC: USIP Press. The best introduction to the logic of conflict resolution is still E. Azar (1990) *The Management of Protracted Social Conflict: Theory and cases*, Aldershot: Dartmouth; or C. Mitchell (2014) *The Nature of Intractable Conflict:*

Resolution in the twenty-first century, London: Palgrave Macmillan. On the subject matter of much of this chapter see O. Ramsbotham, T. Woodhouse and H. Miall (2005) *Contemporary Conflict Resolution: The prevention, management and transformation of deadly conflicts*, 2nd edition, Cambridge: Polity. An excellent summary of the aims of peacebuilding can be found in H.-W. Jeong (2005) *Peacebuilding in Post-Conflict Societies: Strategy and process*, Boulder, CO and London: Lynne Rienner; and H.-W. Jeong (2008) *Understanding Conflict and Conflict Analysis*, London: Sage; and J.-P. Lederach (1997) *Building Peace: Sustainable reconciliation in divided societies*, Washington, DC: United States Institute of Peace Press. A good discussion of the causes and potential resolution of conflicts can be found in A. Oberschall (2007) *Conflict and Peace Building in Divided Societies: Responses to ethnic violence*, London: Routledge. One critical review of conflict resolution approaches can be found in V. Jabri (2007) *War and the Transformation of Global Politics (Rethinking Peace and Conflict Studies)*, London: Palgrave Macmillan. The best recent critique of 'liberal' peacebuilding is by E. Newman, R. Paris and O. Richmond (2009) *New Perspectives on Liberal Peacebuilding*, Tokyo: United Nations Press. Two good overviews of reconciliation after wars are: A. Rigby (2001) *Justice and Reconciliation: After the violence*, Boulder, CO: Lynne Rienner; and D. J. Whittaker (1999) *Conflict and Reconciliation in the Contemporary World*, London: Routledge. On conflict transformation see D. Francis (2002) *People, Peace and Power: Conflict transformation in action*, London: Pluto.

Useful websites

On conflicts themselves, see the unsurpassed series that has been produced by the International Crisis Group since 1995: http://www.crisis-group.org. The best website on conflict resolution/transformation is Fischer, M. and Schmelzle, B. (eds), *The Berghof Handbook for Conflict Transformation* (http://www.berghof-handbook.net/). Various websites related to particular commissions and tribunals mentioned earlier include: South African TRC: http://www.doj.gov.za/trc/; for Rwanda, the official government website: The International Criminal Tribunal for Rwanda (http://www.unictr.org/); http://www.ictj.org/our-work/regions-and-countries/sierra-leone (Sierra Leone). The International Commission on Transitional Justice has the most comprehensive remit of an INGO for such matters across the world: http://www.ictj.org/. The International

Criminal Court can be found at http://www.icc-cpi.int. For the wars in the Former Yugoslavia see: Humanitarian Law Center: http://www.hlc-rdc.org/?lang=de and 'Balkan Transitional Justice' (Balkan Insight): http://www.balkaninsight.com/en/page/balkan-transitional-justice-home.

References

Allen Nan, S., Mampilly, Z., Bartoli, A. (eds) (2012) *Peacemaking: From practice to theory*. Santa Barbara, CA: Praeger.

Angell, N. (1910, 1912) *The Great Illusion: A study of the relation of military power in nations to their economic and social advantage*. London: William Heinemann.

Avruch, K. (2003) Context and pretext in conflict resolution. *Journal of Dispute Resolution* 2: 353–65.

Azar, E. (1990) *The Management of Protracted Social Conflict: Theory and cases*. Aldershot: Dartmouth.

Baker, C. (2015) *The Yugoslav Wars of the 1990s (Studies in European History)*. London: Palgrave Macmillan.

Balcells, L. (2010) Rivalry and revenge: violence against civilians in conventional civil wars. *International Studies Quarterly* 54(2), June: 291–313.

Balcells, L. (2011) Continuation of politics by two means: direct and indirect violence in civil war. *Journal of Conflict Resolution* 55(3), June: 397–422.

Barr, J. (2012) *A Line in the Sand: Britain, France and the struggle that shaped the Middle East*. London: Simon & Schuster.

Bass, G.J. (2000) *Stay the Hand of Vengeance: The politics of war crimes tribunals*. Princeton, NJ: Princeton University Press.

Beevor, A. (2002) *Berlin: The downfall, 1945*. London: Penguin.

Bell, D. (ed.) (2006) *Memory, Trauma and World Politics: Reflections on the relationship between past and present*. Basingstoke: Palgrave Macmillan, pp. 116–34.

Bell, C., Campbell, C., Ní Aoláin, F. (2007). Transitional Justice: (re)conceptualizing the field. *International Journal of Law in Context* 3(2): 81–8.

Bellamy, A. (2012). *Massacres and Morality: Mass atrocities in an age of civilian immunity*. Oxford: Oxford University Press.

Bercovitch, J. and Fretter, J. (2004) *Regional Guide to International Conflict and Management from 1945 to 2003*. Washington, DC: Congressional Quarterly.

Bercovitch, J. and Rubin, J.Z. (1992) *Mediation in International Relations: Multiple approaches to conflict management*. London: Macmillan.

Bew, P. (2009) *Ireland: The politics of enmity 1789–2006*. Oxford: Oxford University Press.

Bleiker, R. (2007) On the Use and Abuse of Korea's past: An Inquiry into History Teaching and Reconciliation. In Cole, E.A. (ed.) *Teaching the Violent Past: History, education and reconciliation*. Lanham, MD: Rowman and Littlefield.

Booth, K. and Wheeler, N. (2007) *The Security Dilemma. Fear, cooperation, and trust in world politics*. London: Palgrave Macmillan.

Browning, C. (2005) *The Origins Of The Final Solution: The evolution of Nazi Jewish policy September 1939–March 1942*. London: Arrow.

Brubaker, R. and Laitin, D.D. (1998) Ethnic and nationalist violence. *Annual Review of Sociology* 24: 423–54.

Bunzl, J. (2008) Mirror Images; perception and interest in the Israel/Palestine conflict. *Palestine–Israel Journal* 12(2–3): 8–14.

Burton, J. (1979) *Deviance, Terrorism and War: The process of solving unsolved social and political problems*. Oxford: Martin Robertson.

Burton, J. (1990) *Conflict Resolution and Prevention*. New York: St Martin's Press.

Burton, J. (1997) *Violence Explained: The sources of conflict, violence and crime and their prevention*. Manchester: Manchester University Press.

Campbell, D. (1998a) *National Deconstruction: Violence, identity and justice in Bosnia*. Minneapolis, MN: University of Minnesota Press.

Campbell, D. (1998b) *Writing Security: United States foreign policy and the politics of identity*. Manchester: University of Minnesota Press/Manchester University Press.

Chandler, D. (2013) Peacebuilding and the politics of non-linearity: rethinking 'hidden' agency and 'resistance'. *Peacebuilding* 1(1): 17–32.

Christie, K. (2000) *The South African Truth Commission*. London: Macmillan.

Clausewitz, C. Von, Heuser, B., Howard, M., Paret, P. (2012) *On War*. CreateSpace independent publishing platform, Amazon.

Coker, C. (1997) How wars end. *Millennium; Journal of International Studies* 26(3): 615–29.

Cole, E.A. (ed.) (2007) *Teaching the Violent Past: History, education and reconciliation*. Plymouth: Rowman and Littlefield.

Corbin, J. (1994) *Gaza First: The secret channel to peace between Israel and the PLO*. London: Bloomsbury.

Corey, A. and Joireman, S.F. (2004) Retributive justice: The Gacaca courts in Rwanda. *African Affairs* 103: 73–89.

Crocker, C.A., Hampson, F.O., Aall, P. (eds) (2007) *Leashing the Dogs of War: Conflict management in a divided world*. Washington, DC: United States Institute of Peace Press.

Dallaire, R. (2005) *Shake Hands With The Devil: The failure of humanity in Rwanda*. London: Arrow.

Darby, J. (ed.) (2006) *Violence and Reconstruction*. Notre Dame, IN: University of Notre Dame Press.

De Tocqueville, A. (1856) *The Old Regime and the Revolution*. Chicago, IL: University of Chicago Press. [*L'ancien régime et la révolution*, 1856]

De Zeeuw, J. and Kumar, K. (eds) (2006) *Promoting Democracy in Postconflict Societies*. Boulder, CO: Lynne Rienner.

Devakumar, D., Birch, M., Osrin, D., Sondorp, E. and Wells, J. (2014) The intergenerational effects of war on the health of children. *BMC Medicine* (12) 57: 15.

Donia, R.J. (2014) *Radovan Karadzic: Architect of the Bosnian Genocide*. Cambridge: Cambridge University Press.

Dragojevic, M. (2013) Memory and identity: inter-generational narratives of violence among refugees in Serbia. *Nationalities Papers: The Journal of Nationalism and Ethnicity* 41(6): 1065–82.

Duffield, M. (2007) *Development, Security and Unending War: Governing the world of peoples*. Cambridge: Polity.

Enloe, C. (1993) *The Morning After: Sexual politics at the end of the Cold War*. Berkeley, CA: University of California Press.

Ferriter, D. (2015) *A Nation and not a Rabble: The Irish Revolution 1913–23*. London: Profile.

Fierke, K.M. (2013) *Political Self Sacrifice: Agency, body and emotion in international relations*. Cambridge: Cambridge University Press.

Fisher, R. (1997) *Interactive Conflict Resolution*. Syracuse, NY: Syracuse University Press.

Fisher, R. (ed.) (2005) *Paving the Way: Contributions of interactive conflict resolution to peacemaking*. Lanham, MD: Lexington.

Fox, J. (2013) Religious armed conflict and discrimination in the Middle East and North Africa: an introduction. *Civil Wars* (15)4, December.

Francis, D. (2002) *People, Peace and Power: Conflict transformation in action.* London: Pluto.

Friedlander, S. (2008) *Nazi Germany and the Jews: The years of extermination: 1939–1945.* London: Weidenfeld and Nicolson.

Fromkin, D. (2000) *A Peace to End All Peace: The fall of the Ottoman Empire and the creation of the modern Middle East.* London: Phoenix.

Fukuyama, F. (1992) *The End of History and the Last Man.* New York: Free Press.

Futamura, M. (2008) *War Crimes Tribunals and Transitional Justice: The Tokyo Trial and the Nuremberg legacy.* London: Routledge.

Gawerc, M. (2006) Peace-building: theoretical and concrete perspectives. *Peace and Change* 31(4): 435–78.

Gray, C. (2005) *Another Bloody Century: Future warfare.* London: Weidenfeld and Nicholson.

Gready, P. (2011) *The Era of Transitional Justice: The aftermath of the Truth and Reconciliation Commission in South Africa and beyond.* New York: Routledge.

Gurr, T. and Harff, B. (1994) *Ethnic Conflict in World Politics.* Boulder, CO: Westview Press.

Hayner, P. (2002) *Unspeakable Truths: Facing the challenge of truth commissions.* New York: Routledge.

Hosoya, C. and A50 Editorial Committee (eds) (2001) *Japan and the United States – Fifty Years of Partnership.* Tokyo: The Japan Times.

Howard, M. (2000) *The Invention of Peace: Reflections on war and international order.* New Haven, CT: Yale University Press.

Huntington, S. (1993) *The Third Wave: Democratization in the late twentieth century.* Oklahoma, OK: University of Oklahoma Press.

Huntington, S. (1996) *The Clash of Civilizations and the Re-making of World Order.* New York: Simon & Schuster.

International Crisis Group Africa (2014) Curbing Violence in Nigeria (II): The Boko Haram Insurgency. Report 216, April.

International Institute for Strategic Studies (2015a) *Armed Conflict Database.* London, IISS.

International Institute for Strategic Studies (2015b) *The Military Balance.* London, IISS.

Jabri, V. (2007) *War and the Transformation of Global Politics.* London: Palgrave Macmillan.

Jabri, V. (2013) Peacebuilding, the local and the international: a colonial or a postcolonial rationality? *Peacebuilding* 1(1): 3–16.

Jeffery, R. and Kim, H.J. (2014) *Transitional Justice in the Asia-Pacific.* Cambridge: Cambridge University Press.

Jeong, H.-W. (2005) *Peacebuilding in Post-Conflict Societies: Strategy and process.* Boulder, CO and London: Lynne Rienner.

Jeong, H.-W. (2008) *Understanding Conflict and Conflict Analysis.* London: Sage.

Jones, D. (1999) *Cosmopolitan Mediation? Conflict resolution and the Oslo Accords.* Manchester: Manchester University Press.

Kaldor, M. (2012) *New and Old Wars: Organized violence in a global era*, 2nd and 3rd edn. Cambridge: Polity.

Kalyvas, S. (2001) 'New' and 'old' civil wars – a valid distinction? *World Politics* 54(1): 99–118.

Kalyvas, S. (2006) *The Logic of Violence in Civil War.* Cambridge: Cambridge University Press.

Kaufman, S.J. (2001) *Modern Hatreds: The symbolic politics of ethnic war.* Ithaca, NY: Cornell University Press.

Kaysen, C. (1990) Is war obsolete? A review essay. *International Security* (14) 4: 42–64.

Keashley, L. and Fisher, R. (1990) Towards a contingency approach to third-party intervention in regional conflict: a Cyprus illustration. *International Journal* 45(2): 425–53.

Kelman, H.C. (2005) Interactive Problem Solving in the Israeli–Palestinian Case: Past Contributions and Present Challenges. In Fisher, R. (ed.) *Paving the Way: Contributions of interactive conflict resolution to peacemaking*. Lanham, MD: Lexington: 41–63.

Krog, A. (1998) *Country of my Skull: Guilt, sorrow and the limits of forgiveness in the new South Africa*. Johannesburg: Random House.

Laffan, M. (1999) *The Resurrection of Ireland: The Sinn Féin Party, 1916–1923*. Cambridge: Cambridge University Press.

Lambourne, Wendy (2009). Transitional justice and peacebuilding after mass violence. *The International Journal of Transitional Justice* 3:28–48.

Lederach, J.P. (1997) *Building Peace: Sustainable reconciliation in divided societies*. Washington, DC: United States Institute of Peace Press.

Lewis, J. (2007) Nasty, brutish and in shorts? British colonial rule, violence and the historians of Mau Mau. *Round Table: The Commonwealth Journal of International Affairs* 96(389): 201–23.

Little, D. (ed.) (2007) *Peacemakers in Action: Profiles of religion in conflict resolution*. New York: Cambridge University Press.

Long W.J. and Brecke, P. (2003) *War and Reconciliation: Reason and emotion in conflict resolution*. Cambridge, MA: MIT Press.

Mac Ginty, R. and Williams, A. (eds) (2007) Commemoration and remembrance in the Commonwealth. *Round Table: The Commonwealth Journal of International Affairs* 96(393).

Mac Ginty, R. and Williams, A. (2012) *Conflict and Development* [in Japanese]. Tokyo: Kuniaki Asomura.

Mack, A. (2006) *Human Security Brief 2006*. Vancouver: Human Security Center, University of British Columbia. Accessed at www.humansecuritybrief.info (accessed on 13 March 2008 but no longer available).

Mandelbaum, M. (1998) *Is major war obsolete?* Survival 40(4): 20–38.

Mearsheimer, J.J. (2001) *The Tragedy of Great Power Politics*. New York: Norton.

Melvern, L. (2006) *Conspiracy to Murder: The Rwandan Genocide*. London: Verso.

Miall, H., Ramsbotham, O., Woodhouse, T. (1999) *Contemporary Conflict Resolution*, 1st edn. London: Polity.

Mitchell, C. (1981) *The Structure of International Conflict*. London: Macmillan.

Mitchell, C. (2014) *The Nature of Intractable Conflict: Resolution in the twenty-first century*. London: Palgrave Macmillan.

Mitchell, C. and Webb, K. (eds) (1988) *New Approaches to International Mediation*. New York: Greenwood Press.

Miyoshi Jager, S. and Mitter, R. (2007) *Ruptured Histories: War, memory and the post-Cold War in Asia*. Cambridge, MA: Harvard University Press.

Mkutu, K. (2008) Disarmament in Karamoja, Northern Uganda: is this a solution for localised violent inter and intra-communal conflict? *Round Table: The Commonwealth Journal of International Affairs* 97(394): 99–120.

Moghalu, K. (2005) *Rwanda's Genocide: The politics of global justice*. London: Palgrave.

Moore, M. (2001) Empowerment at last? *Journal of International Development* 13(3): 321–9.

Newman, E. (2004) The 'new wars' debate: a historical perspective is needed. *Security Dialogue* 35(2): 173–89.

Newman, E. (2013) The violence of statebuilding in historical perspective: implications for peacebuilding. *Peacebuilding* 1(1): 141–58.

Nordstrom, C. (2004) *Shadows of War: Violence, power, and international profiteering in the twenty-first century*. Berkeley, CA: University of California Press.

Nye, J. (2005) *Soft Power: The means to success in world politics*. New York: PublicAffairs.

Oberschall, A. (2007) *Conflict and Peace Building in Divided Societies: Responses to ethnic violence*. London: Routledge.

OECD (2007) *Handbook on 'Security Sector Reform'*. Paris: OECD.

Olsen, T.D., Payne, L.A, Reiter, A.G. (2010) Transitional justice in the world, 1970–2007: insights from a new dataset. *Journal of Peace Research* 47(6): 803–9.

Paris, R. (2004) *At War's End: Building peace after civil conflict*. Cambridge: Cambridge University Press.

Pouligny, B., Chesterman S., Schnabel, A. (2007) *After Mass Crime: Rebuilding states and communities*. Tokyo: United Nations Press.

Prunier, G. (1995) *The Rwanda Crisis: History of a genocide*. New York: Columbia University Press.

Pugh, M. and Cooper, N. with Goodhand, J. (2004) *War Economies in a Regional Context: Challenges of transformation*. Boulder, CO: Lynne Rienner.

Ramsbotham, O., Woodhouse, T., Miall, H. (2005) *Contemporary Conflict Resolution*, 2nd edn. London: Polity.

Ramsbotham, O., Woodhouse, T., Miall, H. (2011) *Contemporary Conflict Resolution: The prevention, management and transformation of deadly conflicts*, 3rd edn. Cambridge: Polity.

Rees, S. (2012) *Passion for Peace*. Sydney: UNSW Press.

Richmond, O. (2006) The problem of peace: understanding the 'liberal peace'. *Conflict, Security and Development* 6(3): 291–314.

Rigby, A. (2001) *Justice and Reconciliation: After the violence*. Boulder, CO: Lynne Rienner.

Roberts, A. (2009) Doctrine and reality in Afghanistan. *Survival* 51(1): 29–40.

Ron, A. (2009) Peace negotiations and peace talks: the peace process in the public sphere. *International Journal of Peace Studies* 14(1) Spring/Summer: 1–16.

Rotberg, R.I. and Thompson, D. (eds) (2000) *Truth v. Justice: The morality of truth commissions*. Princeton, NJ: Princeton University Press.

Ruthven, M. (2003) Edward Said: controversial literary critic and bold advocate of the Palestinian cause in America. *Guardian*, 26 September.

Sandal, N.A. (2011) Religious actors as epistemic communities in conflict transformation: the cases of South Africa and Northern Ireland. *Review of International Studies* 37(3): 929–49.

Sandole, D. (1999) *Capturing the Complexity of Conflict: Dealing with violent ethnic conflicts of the post-Cold War era*. New York: Frances Pinter.

Sandole, D., Byrne, S., Sandole-Saroste, I., Seheni, J. (2009) *Handbook of Conflict Analysis and Resolution*. Abingdon: Routledge.

Sands, P. (2003) *From Nuremberg to The Hague: The future of international criminal justice*. Cambridge: Cambridge University Press.

Tenenbaum, C. (2011) Mediation by Inter-governmental Organisations. In Devin, G. (ed.) *Making Peace: The contribution of international institutions*. New York: Palgrave Macmillan, pp. 67–92.

Thornhill, R. and Palmer, C.T. (2000) *A Natural History of Rape: Biological bases of sexual coercion*. Cambridge, MA: MIT Press.

Tilly, C. (1985) War Making and State Making as Organized Crime. In Evans, P., Rueschemeyer, O. and Skocpal, T. (eds) *Bringing the State Back In*. Cambridge: Cambridge University Press, pp. 169–91.

Touval, S. (1982) *The Peace Brokers: Mediators in the Arab–Israeli conflict, 1948–1979*. Princeton, NJ: Princeton University Press.

Ulrichsen, K.C. (2014) *The First World War in the Middle East*. London: Hurst.

Vayrynen, T. (2001) *Culture and International Conflict Resolution*. Manchester: Manchester University Press.

Wallensteen, P. (2012) *Understanding Conflict Resolution*, 3rd edn. London: Sage (NB the 2nd edition, 2007 is also quoted here).

Walsh, M. (2015) *Bitter Freedom: Ireland in a revolutionary world 1918–1923*. London: Faber and Faber.

Waltz, K. (2001) *Man, the State and War*. Columbia, NY: Columbia University Press.

Webb, K. (1978) *The Growth of Nationalism in Scotland*. London: Harmondsworth.

Webb, K., Walters, M., Koutrakou, V. (1996) The Yugoslavian Conflict, European Mediation and the Contingency Model. In Bercovitch, J. (ed.) *Resolving International Conflicts – The Theory and Practice of Mediation*. Boulder, CO: Lynne Rienner, pp. 171–89.

Wepundi, M., Nthiga, E., Kabuu E., Murray, R., Alvazzi del Frate, A. (2012) *Availability of Small Arms and Perceptions of Security in Kenya: An assessment*. Geneva: Small Arms Survey.

Werbner, R. and Ranger, T. (eds) (1996) *Postcolonial Identities in Africa*. London: Zed.

Whittaker, D.J. (1999) *Conflict and Reconciliation in the Contemporary World*. London: Routledge.

Wierda, M. (2006) Transitional Justice in Sierra Leone. In de Zeeuw, J. and Kumar, K. (eds.) *Promoting Democracy in Postconflict Societies*. Boulder, CO: Lynne Rienner, pp. 183–207.

Williams, A. (2005) 'Reconstruction' before the Marshall Plan. *Review of International Studies* 31: 541–58.

Williams, A. (2006) *Liberalism and War: The victors and the vanquished*. London: Routledge.

Williams, A. (2007) Reconstruction: the bringing of peace and plenty or occult imperialism? *Global Society: Journal of Interdisciplinary International Relations* 21(4): 539–51.

Williams, A., Hadfield, A., Rofe, S. (2012) *International History and International Relations*. London: Routledge.

Winter, J. (1995) *Sites of Memory, Sites of Mourning: The Great War in European cultural history*. Cambridge: Cambridge University Press.

Winter, J. (2000) The generation of memory: reflections on the 'Memory Book'. Contemporary historical studies. *German Historical Institute Bulletin* Fall: 69–92.

Yoshida, T. (2007) Advancing or Obstructing Reconciliation: Changes in History Education and Disputes over History Textbooks in Japan. In Cole, E.A. (ed.) *Teaching the Violent Past: History, education and reconciliation*. Lanham, MD: Rowman and Littlefield, pp. 51–79.

Zartman, I.W. and Rasmussen, J.L. (eds) (1997) *Peacemaking in International Conflict: Methods and techniques*. Washington, DC: United States Institute of Peace.

⑤ Post-conflict reconstruction, democratization and development

Introduction

This chapter will analyse the often-made assumptions about reconstruction, and its allied concepts, democratization, state- and nation-
building, as well as disarmament, demobilization and reintegration
(DDR) especially as they apply to post-conflict situations in developing countries. It will first look at a number of what might be termed
'generic' reconstruction attempts, starting with those of Germany
and Japan (Japan in 1945 was still a 'developing country' and both
operations are even now seen as ideal types), as well as Afghanistan
and Iraq. It will ask whether the remedies proposed under the above
headings can provide the means of what the United States Institute
for Peace called a need for 'Managing Global Chaos' (Crocker
et al. 1996).

The period since the end of the Cold War has not been characterized
by a linear rise in intra-state wars. There was an important rise in
1991–2, but a decline in such conflicts until 2007 (Peace and Conflict
website, University of Uppsala, www.pcr.uu.se). From the perspective
of intervening powers and IGOs these wars are often termed 'small
wars', in that they require, after an initial large investment of troops
and material (as in Afghanistan in 2001 and Iraq in 2003), relatively
small numbers of troops, often in counter-insurgency (COIN) mode.
The forceful peacemaking efforts of the 'coalitions' in Afghanistan
and Iraq are widely perceived to have been ineffective, even when
conducted by troops with a great deal of experience of these types of
operation, especially in Northern Ireland and in post-imperial operations, like the British army (Ledwidge 2011, but see for a contrary
view Sky 2015). There have been some post-Cold War cases that are
seen as 'successes', notably in Sierra Leone and Kosovo (both 1999).
But these have tended to be eclipsed by what has happened since,

especially in the Middle East, where there has been (as of mid-2015) a conspicuous lack of international intervention in the Libyan and Syrian civil wars. Even COIN is now seen as ineffective by some influential commentators (Etzioni 2015). Non-intervention (save for aerial bombardment) has now seemingly become the default position for Western states like Britain, France and the US, compared to the gung-ho intervention of the early 2000s.

So what has gone wrong? How can we square the enormous resources deployed in wars in developing countries with their relative failures? One way is to look at what might be termed the changing 'semantics' of reconstruction, for it is in the terms of a phenomenon's description that its core values can usually be determined. In turn, a presentation of a term can lead to its being seen as a use of power, positive and negative. To say as much is not to be 'anti-American' or to accuse the Western Powers of 'imperialism', but to identify how these actions can be seen as such, no matter how pure the intentions of the intervening Powers or even how effective their actions in helping the cause of long-term peace (Williams 2007). For a casual reader of the literature on how we should go about bringing peace to war-torn societies after, usually, civil wars is presented with a plethora of seemingly contradictory terms. Our aim is to show how the ideal of 'peacebuilding' described in Chapter 4 relates to 'reconstruction' in the 1990s and since, as well as to 'state-' or 'nationbuilding'.

Much of a chapter like this will use terms that are often conflated, although in reality they have had many different meanings histori-cally. Shortly we will look at how that applies to the key term 'recon-struction'. 'Nation-' and 'state'building have similarly gone through many different iterations. They are often conflated, and both usually figure in defences of 'peacebuilding' efforts. But of course they do have distinct implications. Many 'nations' do not have 'states'. We could see a nation as a group of people who identify themselves as a coherent current, as well as historical, and usually linguistic grouping like the Kurds, or the Kikuyu in Africa. But not all such groups have, or indeed wish to have, a 'state'. Hence studies of nationalism often differentiate between 'civic' and 'ethnic' national identity. Many states are multinational, like the United Kingdom. In the pages that follow we will attempt to show how the two concepts of 'nation' and 'state' can and have been used or abused in the context of policies of intervention and reconstruction.

Changing definitions, changing historical contexts

There are many definitions of what we would now loosely call 'reconstruction'. We use a broader definition than that usually employed. Hence, we believe that:

> [R]econstruction encompasses short-term relief and longer-term development. It extends far beyond physical reconstruction to include the provision of livelihoods, the introduction of new or reformed types of governance, and the repairing of fractured societal relationships. Thus reconstruction is not merely a technocratic exercise of rebuilding shattered infrastructure. Instead, it is an acutely political activity with the potential to effect profound social and cultural change. Post-war reconstruction holds the capacity to remodel the nature of interaction between the citizen and the state, the citizen and public goods, and the citizen and the market.
>
> (Mac Ginty 2007: 458)

Even this catch-all definition has its problems of course. It leaves hanging what kinds of 'interactions' are envisaged, and between whom. We hope to show that the usual default *model* that is most often used in statebuilding and reconstruction attempts is that of the 1940s, and the reconstruction of Germany and Japan using the twin levers of military domination and economic mobilization through the Marshall Plan. So what if this model is fundamentally flawed when we attempt to deal with civil war situations of the post-war period that do not lend themselves to comparison to those of the 1940s?

The term 'reconstruction' has two main semantic roots, in the logic of colonial/imperial administration and in the American Civil War (Foner 1989; Williams 2005; Cramer 2006).

The first of these has maybe had the longest impact on thinking about the practice in developing countries and the erstwhile colonial, and especially the European Powers. In the nineteenth century, the idea was used a great deal by great liberal thinkers like John Stuart Mill to convey the need to 'civilize' the 'natives', as Mill put it under the guidance of 'philosophical legislators' to create a 'Greater Britain' through colonial emigration to what are now developing countries in the Commonwealth (Bell 2007: 50). Unfortunately, the implantation of white farmers, which was often linked into 'punitive expeditions' to clear 'natives' from large areas of Africa, has had the long-term result of causing massive resentment that came out in such conflicts as the Mau Mau in Kenya in the 1950s and in inter-tribal violence as a sequel to colonially imposed 'solutions' to land and other problems (Lewis 2007). As the distinguished British political theorist John Plamenatz explained at the end of the colonial period, in 1960: '[n]o western people have cared more for freedom than the British and the French... and to elaborate the rules and institutions needed to establish it. They have also created the largest empires in Asia and Africa... True, they did not create them to bring freedom to others so much as to increase their own wealth and power' (Plamenatz 1960: 17). Many of the issues now facing countries like Kenya, Zimbabwe and South Africa were in part caused by the unintended consequences of liberal attempts to create modern political and economic state structures, which are usually seen by indigenous peoples as being part of a system of institutionalized inequity.

In the aftermath of the Second World War the term 'reconstruction' became synonymous with the emergence of the first United Nations agency, the United Nations Relief and Rehabilitation Agency (UNRRA) in 1943–47, and the Marshall Plan in 1946–52, which did much to help resettle refugee and former concentration camp populations and rebuild much of the Western European and Japanese economies in the largest such enterprise to date (Hogan 1987; Williams 2005a and 2006). The creation of the international financial architecture (the International Monetary Fund and the World Bank in particular) also did much to establish the context in which German and Japanese reconstruction could take place. Hence macro-level reconstruction (of the international system) enables micro-level reconstruction at the country level (for more on this see Box 5.1).

International governance of territories was also given an IGO imprimatur, as with League of Nations rule over the Free City of Danzig under the provisions of the 1919 Treaty of Versailles from 1919 to 1939, and the Saar from 1922 to 1935. In several ways we could see this form of activity as having had the same effect as with UNRRA in helping improve the normative image of reconstruction efforts in general. The League's administrators undoubtedly did much to protect minority populations, to act to deal with disputes between the territories and other nearby states and to develop constitutional guarantees for the population. This has continued to be seen as a successful strategy since the end of the Cold War by the EU, OSCE (Organization for Security and Cooperation in Europe) and UN among other IGOs in Cambodia (UNTAC, 1992–3), Bosnia-Herzegovina (1994–present), Kosovo (UNMIK, 1999–2008) and East Timor (UNTAET, 1999–2002) (Caplan 2005: 29 and Chapter 1).

Box 5.1

German and Japanese reconstruction, 1945–55

At the end of the Second World War the victorious Allies (the 'United Nations', led by Britain, China, France, the USA and the USSR) decided to dismember Germany into 'zones', dismantle the political, economic and social organizational structure of the National Socialist State and in effect rebuild the German nation. The same fundamental logic was used for Japan and, to a much lesser extent, Italy. The process was interrupted after 1947 by the onset of the Cold War between the former partners, but the Western zones of Germany, Japan and Italy all benefited from a huge ($50 billion) injection of soft loans and direct aid to rebuild the shattered infrastructure of the countries and more widely of Western Europe as a whole. This 'Marshall Plan' was largely successful in creating or at least enabling the 'economic miracles' in Germany and Japan in the 1960s. There was little or no obstruction from the previous power hierarchies who had been comprehensively defeated in battle and were subject to awful privation and violence on a continent brutalized by total war. Reconstruction took place on the following levels: physical infrastructure; institutions; political/cultural – 'de-Nazification' or 'democratization'; legal; military, and; civil and economic. The legal efforts aimed at 'reconciling' the defeated countries to their neighbours both internally and by a form of retributive justice (see pp. 149–51) in war crimes tribunals, some of which were directed by international legal teams.

Sources: Lowe (2013), Beschloss (2002), Dobbins, McGinn et al. (2003), Killick (1997).

These cases have become paradigmatic in thinking about nationbuilding and reconstruction ever since. To summarize, the reconstruction exemplars of Germany and Japan have to be seen in the context of a 'reconstructed' macro-international system (especially its public and private financial architecture), and a clear and enforced change in not only the political, but also the cultural (including economic) mindsets of the countries involved. It was made clear to their populations that only certain beliefs and activities were acceptable and others deviant. It might be argued that this logic has been applied to 'developed' as well as 'developing' countries since 1945. The tools are those of 'conditionality' for grants and soft loans, often imposed by the IMF, as was done to Britain in the 1970s and many LDCs. It might even be said to have been demonstrated by the way the EU Commission and the European Central Bank have insisted on 'conditionality' for loans given to Greece, Ireland and Portugal since 2009. But one evident example is that of Argentina. Argentina was declared bankrupt in 2002, leaving about eight million of the 35 million population 'destitute' (United Nations Economic Commission for Latin America and the Caribbean) (ECLAC 2002). It is possible to avoid the logic of such IMF conditionality, but it will result in economic misery and this is naturally greatly resented by subject populations. Social unrest and a distrust of IGOs and national governments is an almost inevitable result.

Reconstruction after the end of the Cold War

There are important differences to the normative background to discussions of efforts at 'reconstruction' in the post-Cold War period. Primary among these is a changed environment for claims to, and discussions about, sovereignty, a subject we broached briefly at the start of the last chapter. Sovereignty has been termed the 'master noun' of international relations and derives much of its underlying meaning from classical Western political thought, and especially from thinkers like Thomas Hobbes and Jean Bodin. This has consequently greatly coloured the global view of what a state is and what rights and duties states owe their citizens and vice versa (Slomp 2008). It might be argued that, as with the debate on democracy, the Western Powers that dominate the global system have come to be seen in developing countries as moving the goalposts whenever developing countries interpret sovereignty in ways deemed unacceptable in Western capitals. So, as many writers (e.g. Caplan 2005; Jackson 2005) have pointed out, the 'new interventionism' and its allied concept of a 'responsibility to protect' (R2P) represent both a realist, national-interest-based justification for intervention to address UN Charter Chapter VII 'threats to international peace and security' and liberal impulses to protect vulnerable populations in civil wars. Governments and populations in developing countries may be wrong to see R2P as 'imperialism' on occasions, as in Afghanistan, Iraq or Sudan currently, but they nonetheless do often consider that to be the case (Bellamy 2014; Hehir 2012).

As a consequence of these twin impulses, various new epithets have emerged to join with the existing ideas of reconstruction and international administration that pre-date 1990. Primary among these are 'nationbuilding' and 'statebuilding'. The first of these is usually taken to refer to the 'people' of a given area and to an ethos, the second to the building of institutions, though the two have often been semantically

Plate 11 *Rebuilding a war-damaged bridge in southern Lebanon: post-war reconstruction includes not only infrastructure but also rebuilding fractured relationships in society.*

confused in recent times. Caplan distinguishes 'international administration... from nation- or state-building (the two terms are frequently used interchangeably) though state-building is very often an integral part of an international administration' (Caplan 2005: 3). We might therefore suggest that the definition of what exactly 'reconstruction' means changes in line with the state of the international system and the obsessions of the actors that are prominent power-holders within it, rather than the objective needs of the populations affected by conflict and war. Thus, for example, the term 'resilience' is trendy at the moment but will probably be overtaken by a new word in a few years.

Nationbuilding

The successes of the essentially United States-funded initiatives in the 1940s and the early1950s gave rise to a belief that the 'nationbuilding' of Japan and Germany described above were paradigmatic examples of how it could and should be done and were much evoked in subsequent reconstruction and development efforts, notably, and probably most misleadingly, in the aftermath of the 2003 invasion of Iraq. Francis Fukuyama, an advocate of the intervention in Iraq, is surely right when he says that both conservative (for which read mostly Republican) and Democrat (or 'liberal') Americans 'have come to support nation-building efforts at different times – conservatives as part of the "war on terrorism" and liberals for the sake of humanitarian intervention'. It might also be said that many Americans would defend the parallels of the intervention in Afghanistan and Iraq with that of Germany and Japan and refer to this as 'nationbuilding' – as in both cases they see it as 'constructing political institutions ... coupled with economic development'. In American minds 'nation-building reflect[s] the specifically American experience of constructing a new political order in land of new settlement without deeply rooted peoples, cultures and traditions' (Fukuyama 2006: 1–3). However, this was certainly not the case in Germany, Japan, Afghanistan or Iraq, where there were existing political and other traditions, albeit illiberal ones.

'Nationbuilding' is explicitly defined by the American think tank, the Rand Corporation, as 'the use of armed force as part of a broader effort to promote political and economic reforms with the objective of transforming a society emerging from conflict into one at peace with itself and its neighbors' (Dobbins et al. 2007: xvii). This explicit espousal of military force sits uneasily with those who wish reconstruction to come from within a society, the basic desire of the liberal, and not to be imposed from without, the basic instinct of the realist. The Rand authors call this distinction 'co-option', where 'the local population are actively involved in all stages of the planning and execution of a nationbuilding' exercise and 'deconstruction' as was the case in Germany and Japan in 1945 (Dobbins et al. 2007: xx).

This easily gives rise to claims that the United States has used a relatively benign expression to hide its essential national interest, or what Jacoby calls the '*Realpolitik* of [its] hegemonic interest'. He finds it strange indeed that parallels with the Second World War reconstruction efforts have been used uncritically and ahistorically by the United States' administration to justify policies that are muddle-headed and 'focussing on "how", rather than "why"' they should be undertaken (Jacoby 2007: 522).

But more fundamentally, one of the key problems with the logic deployed since 2001 is the implicit assumption that we can indeed 'build' states. For the key to asking that question is the analogous one of asking can we build 'nations'? Put like that, the answer has to be either that we think that somehow nation-states are somehow 'natural' units of political organization, or that there is a model we can follow that will guarantee the success of any such endeavour.

The main results of 'nation-' or indeed 'statebuilding' have not been happy in many instances. In Rwanda in 1994 we saw the 'genocidal state' (Werbner in Werbner and Ranger 1996: 12–13; Lemarchand 2009). In Zimbabwe the state started in 1979 with a bout of 'ethnic cleansing' around Bulawayo and has continued in like fashion. This is a far cry from the democratic dream of many of the nation's new citizens. So we argue that the emphasis on a particular kind of 'statebuilding' is debatable to say the least, and more likely to be based on European experiences of political organization, not that of other areas of the world where colonial Powers imposed such thinking on subject populations. One of the reasons that colonial Powers like the British and French were forced into what Martin Thomas has termed 'Fight or Flight' is because the local populations did not want European models imposed on them, but that is what they have experienced nonetheless (Thomas 2014).

Equally, one of the issues facing countries emerging from post-colonial and/or civil wars is that the constituent parts of the state have no binding 'identity', in that they often in no way see themselves collectively as a 'nation', in the way that, say, the multi-lingual and multi-confessional Swiss do. Creating that shared identity cannot merely be done by creating common institutions, important as they are.

The most likely result has proved to be a variant, as in Bosnia and Timor-Leste, of international governance. In Bosnia this governance is through the Office of the High Representative (http://www.ohr.int/), who can overrule the decisions of local politicians if they do not fit the desired model. Some authors have suggested that this is the only way to 'remake' successful communities (Toal and Dahlman 2011). Lara J. Nettlefield, in talking about the attempts by the international community to bring justice through the International Criminal Court, is more positive. In a counterblast to the argument that the Bosnian Dayton Accords are merely 'faking democracy' (Chandler 2000), Nettlefield suggests that we should not be so harsh in our judgements about 'pioneering institutions' (Nettlefield 2010: 7).

But Nettlefield also acknowledges that these institutions can only hope to be some solace to those who have lost their 'youth' and 'political community' (Nettlefield 2010: 7). With all the old ones lost, shared 'symbols' have to be created (Keranen 2014) to overcome 'different visions of community and belonging'. Of course this argu- ably takes a very long time indeed to become anchored in the popular consciousness. Other writers question the 'purpose, legitimacy, goals and effectiveness of "nation-building"' and point to its 'mixed' record. The imposition of democracy, a prime aim in all such efforts since 1990, has seemingly brought about a semblance of order (Bosnia, Namibia, Sierra Leone and Timor-Leste are key examples), while in others (like Cambodia) peace, but not democracy, has resulted, and in yet others (Somalia, Afghanistan, Iraq, the DRC, Haiti) neither peace nor democracy has resulted (Hampson and Mendeloff, in Crocker et al. 2007: Chapter 37). Hence the idea that the creation of liberal demo- cratic institutions and the implementation of market 'reforms' (which often means helping an economic elite at the expense of the mass of people) will lead to a more equitable and legitimate division of power is often an illusion. For example, as Meike de Goede has put it in a particularly apt analogy, many people see politicians as 'thugs in ties', or *'Kulana en cravate'*, in the Democratic Republic of Congo (DRC) (de Goede 2015: 2).

So, one clear explanation of the popularity of the term and practice of 'reconstruction' is that it is a term that is deeply embedded in the historical consciousness of the great Western Powers, and especially of the United States of America, and that this is mainly due to the successful use of the idea in the pacification of Germany and Japan after the Second World War. But it has to be underscored that the term has a *different* resonance in other parts of the world, in particular in the developing countries that are our main focus. For them 'recon- struction' and its successor terms, 'nation-' and 'statebuilding', have the tinge of 'imperialism'. What seems like a good idea to a victorious power is not necessarily experienced in the same way by a vanquished one. It also has direct consequences for interveners, as happened in Somalia, for example. To use Keen's expression: '[o]ne of the most important roots of violence is a sense of having been *humiliated*' (Keen 2007: 50, his italics). Many of those the West aims to help through 'reconstruction', and indeed through the institutions that they use to exercise this help, do feel so 'humiliated' (Badie 2014).

It is therefore inescapable to avoid describing the intentions and actions of IGOs and major states alike in places like Afghanistan and Iraq as post-imperial *hubris*. Such has been this *hubris* that since 2003 and the invasion of Iraq, we have seen books emerging with snappy titles like *A Beginner's Guide to Nation-Building*, a kind of do-it-yourself approach to dealing with failed states, produced by the Rand Corporation (Dobbins et al. 2007) with useful graphs and chapters on every conceivable aspect of the task. Such manuals assert that although '[e]ach nation to be rebuilt may be unique … the nation-builder has only a limited range of instruments on which to rely'. The 'instruments' are itemized as various kinds of military and administrative personnel, as well as 'experts in political reform and economic development' (Dobbins et al. 2007: vii).

This process tends to exclude the 'locals' at the expense of outside 'experts', and books like that by Dobbins et al. do tend to put such an emphasis on the outside agents concerned. But can the locals be ignored? When they are, they tend to see the institutions created as illegitimate. This in turn has had an effect on the strength of the state, and this can become a problem not just for one 'nation' but for many, creating the kinds of *regional* destabilization that we see in Central Africa. As a consequence there are destructive effects across huge swathes of territory for civil and political norms that states are supposed to uphold – the rule of law, civil society, human rights, and so forth – as well as for social and economic norms such as economic stability, poverty reduction, health and many more besides.

Statebuilding

The difficulties of creating a 'nation', or healing one after a civil war, are just as complicated as building or mending a 'state'. This has been put into even sharper evidence by the events in the Middle East since the beginning of the Arab Spring. The revolution that has broken up the 'states' that were Iraq and Syria has turned the populations into warring factions, has ended the rule of the capitals and governments and created or reinforced new national entities like the Kurds (who now have both an effective 'nation' and 'state') and transnational entities like the 'Islamic State' (also know as ISIS or *Daesh*). Both the Kurds and *Daesh* have more or less effective power over the territories they control, though neither is recognized by the international community, nor by any of the states or IGOs that make it up. Even before or coterminous with the current crisis there were other entities such as 'Transnistria' (a part of Moldova) that were not recognized but operated as 'states'. We have other such territories in what is officially 'Ukraine' that have declared themselves independent ('The Donetsk People's Republic') or been annexed by other states (such as Crimea by Russia). The first example of this after the end of the Cold War was the establishment of Eritrea in 1993. There was only one secession during the Cold War, that of Bangladesh in 1971. No one expected the trend to take root. But this is a phenomenon that is likely to grow, as we have seen with (albeit democratic) movements to create new states of Catalonia and Scotland. Sadly not all of these movements have led to an improvement of the development or human rights records of the new states. Eritrea is one such cautionary tale, where after an optimistic start in the 1990s hundreds of thousands fled war and an authoritarian state in the 2000s, often only to drown in the Mediterranean whilst seeking to reach an increasingly rejectionist Europe, which nevertheless bears at least some responsibility for their fate (Beretekeab 2000; United Nations Human Rights Commission 2015 http://www.ohchr.org/Documents/HRBodies/HRCouncil/CoIEritrea/A-HRC-29-42_en.pdf).

So who or what is to blame for this sorry state of affairs? One culprit has to be the 'international community'. The lack of international official acceptance of the changed geometry of state borders or jurisdictions on the ground has in the past led, for example, to the state of Eritrea being denied official recognition in spite of the evident loss of control of that area by the previous official state, Ethiopia. In that case the capture of Addis Ababa by two (then) allied liberation movements led to a temporary agreement over power-sharing and border delimitation that ended a civil war that had raged for over 30 years in the Horn of Africa. But what if such agreement is not possible, due to the refusal of surrounding states or the international community to accept the new status quo?

Another culprit (if that is the right word) is that in developing areas of the world like Central Africa, the Horn of Africa, as well as in the Former Soviet Union and Yugoslavia, globalization and state failure have combined to create a caste of freelance economic military entrepreneurs. One example is the followers of Foday Sankoh in Sierra Leone in the 1990s, or of Charles Taylor in Liberia, many of them disaffected and very violent young men (Hoffman 2011; Wiafe-Amoako and Mazrui 2014). These groups often claim ethnic or cultural motivations, but also often have distinctly economic motivations, such as the extraction of commodities for their own enrichment (Pugh and Cooper 2004). This combination of appeals to 'greed' or 'grievance'

(Collier 1999) has 'state-disintegrating' consequences, as in West Africa and the DRC, that are difficult to deal with locally given the enormous resources that wars put in the hands of these warlords.

However, it must also now be recognized that many civil wars, such as those in Africa or those now raging in the Middle East, are to do much more with identity issues, sometimes due to pre-modern allegiance to clan or tribe, and often allied to ones of confessional allegiance. These kinds of allegiance are not ones that sit easily with the models of conflict settlement, resolution or management that we discussed in the previous chapter. The idea that 'power sharing' is going to work, as is often suggested (as by Sisk 2013), is not always evident. IGOs and states like to deal with their peers, not with revolutionary movements who want to change borders, defy international conventions, even on such issues as human rights, and who do not accept the legitimacy of IGOs like the UN or its Security Council. In short, these new entities want to subvert the international system as a whole.

Of course this does not mean to say that all intra- or inter-state wars are due to a total rejection of the notion of the 'state', just the state some of the population find themselves in. The section of the population that accepted the annexation of Crimea to Russia was undoubtedly keen to benefit from the enhanced security, economic and identity reinforcement that they perceived Russia being able to provide over that of the 'official' state Ukraine. There may even be those who welcome ISIS as a better guarantor of their basic needs than the official states of Syria and Iraq. Even Charles Taylor and Foday Sankoh promised those who followed them a better life, albeit with such slogans as Taylor's in 1997: 'He killed my ma, he killed my pa, but I will vote for him.' He won the election with 74.3 per cent of the vote (Meredith 2011).

As a further twist in a complicated tale of statebuilding, it could also be said that we need to temper the 'new war' emphasis on the negative economic results of globalization on many wars, as it could also be claimed that economic factors have persuaded some warriors to lay down their arms in return for the economic benefits of peace. This was partly the case in Northern Ireland and, maybe, in the Former Yugoslavia and could be seen as one plausible explanation of the seeming paradox of the reduction of the number of wars discussed at the beginning of this chapter. The salutary effects of economic interdependence on war and peace are much as they were when first described by Norman Angell in 1912 (Angell 1912) and now by scholars who follow a broadly similar logic. But we would still maintain that the logic of economic gain generally tends to reinforce Pugh and Cooper's thesis about economic drivers in civil wars and Carolyn Nordstrom's findings of the essentially economic nature of the motivations of many who fuel the 'new' wars in the context of globalization (Nordstrom 2004).

In many ways, the features of, and the solutions to, these processes are analogous to what happened in Europe in the late Middle Ages when state structures grew and eventually led to the emergence of systems of security and law. Other writers have referred to a 'neo-Medievalism' that characterizes the wars we are now seeing, defined by Hedley Bull as 'a system of overlapping authorities and criss-crossing loyalties' (Bull in Winn 2004). Indeed one of the differences between many developing states and those of established 'developed' ones is that they precisely are often not 'a sovereign entity dominated by a single predominant national culture. An enormously compelling mixture of legitimacy and efficiency.' Instead, much of the developing world more resembles the City States of Renaissance Italy or Europe before 1630 than the settled nations of modern Europe (Winn 2004: 2–3).

But, as Herfried Münkler has pointed out, the difference with the Europe of the Thirty Years' War, which devastated huge areas of Europe between 1618 and 1648, or the English Civil War (1642–9), is that 'the *state-building wars* in Europe or North America ... took place under almost clinical conditions, with no major influences "from outside", whereas this has not been the case with the *state-disintegrating wars* in the Third World or the periphery of the First and Second Worlds' (Münkler 2005: 8). We might also add that although the Westphalian system of states was the ideal after 1648 in many parts of Europe, it was not the reality until very recently. What we are seeing now in the Middle East might also be said to be a return to the instability of states and the potential for transnational movements that wish to recreate not states, but pre-modern entities like the Ottoman Empire, or even a 'Caliphate'. This trend might get worse as the tensions and civil wars that we are now seeing in Nigeria and surrounding states, as well as in the Horn of Africa, lead to analogous demands for local 'caliphates'. A large number of states in Africa have not been characterized by economic or political stability in the past few decades, which might prove to have been a harbinger of their radical transformation or even collapse.

So, how can *state-building* be brought about? As we have seen, the end of the Cold War has led to a very different set of problems than that which prevailed until 1990. Caplan has summed it up as follows:

> State-building refers to efforts to reconstruct, or in some cases to establish for the first time, effective and autonomous structures of governance in a state or territory where no such capacity exists or where it has been seriously eroded.
>
> (Caplan 2005: 3)

The next section will explore how the underlying rationale(s) used in such efforts may be said both to hold out hopes of success, but also to point to inherent dangers.

The 'stages' of peacebuilding and reconstruction: towards 'sustainable' peace?

Many writers on nationbuilding, statebuilding, or peacebuilding see the process of achieving that in somewhat mechanistic terms. So, for example, Roeder and Rothchild (2005) lay down a series of 'stages' that they believe are essential if the international community is to bring 'sustainable' peace to an area. Jeong (2005) sees them as necessarily 'sequential'. The touted solutions usually have to do with the encouragement or imposition of 'democracy' which corresponds with the liberal mantra that 'democracies do not go to war with each other'. Empirically that is true, but the concern is that countries in a state of *transition* to democracy, or even more so in a transition to *statehood*, do. Developing countries are usually in states of major transition of some kind and are consequently more violent than those that are not.

Paris (2004) has suggested that given the difficulties of transition to democracy it would be better to concentrate on 'institutions' before changing political practices. The initial problem in the reconstruction of any developing country, he notes, is the lack of basic security. In successful states, the monopoly of power is held by the state; in states that have experienced a civil war this is certainly not the case. So 'the first task of peacebuilding is to restore the monopoly as a foundation and precondition for all further institution-building efforts' (Paris 2004: 206–7).

We shall therefore now outline how that initial logic usually unfolds in terms of policy prescriptions and suggest how many writers have seen problems in the implementation of those prescriptions.

Actors

The primary actors in the processes of reconstruction/statebuilding since 1990 have been as follows.

States

In spite of George W. Bush's famous dictum 'we don't do nation-building', during a presidential candidates' debate in 2000, that is precisely what the United States has been doing in many parts of the world. The European Union has in many ways been an enthusiastic supporter of such efforts, taking a leading role in the Former Yugoslavia through the European Agency for Reconstruction (http://ec.europa.eu/enlargement/archives/ear/home/default.htm), as well as in Cambodia. The UK has played a determinate role in Sierra Leone with military, civil and other assistance on a large scale, through the innovative Conflict Pools system and the Stabilization Unit pioneered by the Foreign and Commonwealth Office, Department for International Development (DFID) and the Ministry of Defence (Ginifer 2004). So has France in its former colonies, such as Ivory Coast and Chad, often through the African Development Bank (ADB). The EU and other groupings of states, like the Commonwealth, have played an important role in election monitoring. Even Japan has played a rather more minor role (in Cambodia and Iraq: Japan MOFA 2007), one that may now expand with the changing of Japan's constitution by Prime Minister Shinzo Abe in September 2015.

IGOs and IFIs

The role of intergovernmental organizations (IGOs) and international financial institutions (IFIs), like the IMF in Russia in the immediate aftermath of the Cold War, has been followed by a subordinate role for that organization in other places. There has been the establishment of such specialist bodies as the European Bank for Reconstruction and Development (EBRD) and the Organization for Security and Cooperation in Europe (OSCE).

The role of the World Bank has undoubtedly been the most important in reconstruction efforts, providing technical assistance in myriad ways. The way this has been applied has drawn criticism even from within the Bank itself (Stiglitz 2003) for advocating policies that tended to kill the patient, rather than cure it. The World Bank has moved in recent years from the much-criticized Structural Adjustment Programmes (SAPs), which often reduced the social safety nets for the poor in developing countries, in favour of 'Poverty Reduction Strategy Papers' (though neither of these are specifically for post-conflict countries). IFIs also act as important coordinators of effort through such mechanisms as 'donor conferences'. As we have noted, the UN's 'Millennium Development Goals' of the period 2000–15 have now been succeeded by 'Sustainable Development Goals', which have much the same aims of eliminating poverty, but with a new emphasis on 'resilience' and 'sustainability', hopefully a more realistic approach than the first 'Goals' (United Nations Development Programme 2015).

Most such organizations are Western inspired and based, if nominally international. There are comparable economic organizations in the developing world, such as the Economic Community of West African States (ECOWAS), and regional development banks (such as the AfDB). There are also embryonic political organizations, like the African Union, which is playing an important role in Darfur in Sudan. But the really big financial resources still come from Western sources.

NGOs

On the ground, NGOs have played vital roles both in primary assistance, such as the alleviation of hunger and with medical intervention, in war and post-war situations, but also monitoring roles (as with the International Committee of the Red Cross and the Red Crescent). There has long been a tension between the human rights and civilian focus of such groups and the military/security focus of peacekeepers and military personnel more generally in the final stages of a conflict. 'Soft' and 'hard' NGOs and IGOs have on occasion found it difficult to cooperate, as in Iraq before 2003 when there was some friction between human rights NGOs and IGOs engaged in monitoring the (alleged) weapons of mass destruction.

The press

The press presence is a phenomenon that has drawn much attention since the first Gulf War of 1990–1, and conflicts in Somalia, Rwanda and Bosnia in the early 1990s, with the 'CNN' effect, the effect on

Western public opinion of the presence of media reporters in the midst of disaster areas (Hammond 2007: 12–13), being seen as a major pressurizer of governments to be 'seen to be doing something'. This is part of what might be called the 'framing' of international conflicts, affecting the types of intervention undertaken in significant ways (Hammond 2007). The media is also seen as a major need in the creation of civil societies by such bodies as the EU (Loewenberg and Bonde 2007).

Demobilization

After the 'ceasefire', the job has only just begun, a realization that first really dawned at the end of the First and Second World Wars. How do you pick up the pieces left by violent conflict? The job of disarming the German armies in those wars' aftermaths was relatively simple as they were disciplined organizations that obeyed orders and could be largely trusted to comply with agreements. The disorganized militias of the current period have a tendency to split into ever-smaller factions. There are also 'drivers' that often mean holding on to their weapons is a rational choice in a way that an army demobilizing to return to a normal civilian life of work and family would not contemplate.

The international organizations, especially the United Nations, were slow in realizing the importance of disarming insurgent groups in attempts at post-conflict development. Partly this was because in the early 1990s it was not fully appreciated how many wars would now be 'internal'. The first understanding of this came with the *Agenda for Peace* (Boutros-Ghali 1992), but as late as 1997 organizations like the Carnegie Commission (Carnegie 1997) put less emphasis on 'conventional' disarmament than they did on that of nuclear, chemical and biological weaponry. It was not until the UN Brahimi Report of 2000 (Durch et al. 2003) that a clear commitment to conventional disarmament got a real boost and a consequent stressing in policy debates. The United Nations Institute for Disarmament Research (UNIDIR, http://www.unidir.org) has published annual reports on the progress of individual missions by the UN in countries like Cambodia, Sierra Leone and many others. The World Bank has been the leader in demobilization and reintegration projects but did not get involved in disarmament until about 2003 (Muggah 2005: 243–4).

Disarmament, Demobilization and Reintegration

The resulting policy trio, 'Disarmament, Demobilization and Reintegration', often referred to by the acronym 'DDR', is directed at reducing the most obvious proximate source of violence in a conflict (for a full discussion of DDR see Özerdem 2008; McMullin 2013). As Durch et al. (2003: 26) put it: 'DDR programs are essential elements of post-conflict stability that reduce the likelihood of resumed conflict.'

Removing weapons through their collection and subsequent destruction ('Disarmament') is important symbolically, as it signals the parties are willing to pursue political rather than violent means. This is especially so if the process is accompanied by registration and monitoring of all factions and their return to former home districts or barracks ('Demobilization'). In civil wars in developing countries this will quite often require the official state forces to accept a diminution of their role as sole arbiters of control.

However, the main system-level solution is always the 'reintegration' of ex-combatants, which Özerdem describes as 'becoming civilian' (Özerdem 2008). This often, even usually, requires the implementation of training programmes to enable former soldiers who may never have known anything but war to find different kinds of job, though that is often in the 'official' state army or police. It must be remembered that many wars in developing countries have been fought by children, notably in Central and West Africa (Beah 2008; Özerdem and Podder 2011). For a detailed analysis of the reintegration problems in Liberia see Podder (2012). But to provide a satisfactory alternative to the profession of arms is notoriously difficult and former child soldiers often have to be 'reintegrated' into populations that were previously their victims. Education also proves difficult for young people who are themselves traumatized by what they have seen and done (Betancourt et al. 2008).

The stated aim of reintegration is to help create a working democracy. This includes reviewing the civil/military relationship, a feature in many coup-prone countries (as with Fiji or Pakistan), or with countries where dictators have an unhealthy relationship with their armed forces, as for example in Zimbabwe, which is arguably worse in the long run for local and regional prospects for peace and prosperity alike. An important corollary to this is 'Security Sector Reform' (SSR), where the army and police are re-formed to make them more

reflective of democratic norms and practices, not mere protectors of
the elite and their own livelihoods. SSR has been defined as follows:
'Security sector reform aims to develop a secure environment based
on development, rule of law, good governance and local ownership of
security actors' (GFN-SSR http://www.ssrnetwork.net/about/what_is_
ss.php, accessed 14 July 2015).

These efforts can collectively be summed up as attempting

> [t]o…devise transitional arrangements for the short term. These must
> provide the modicum of political stability necessary to conduct elec-
> tions to a constitutional assembly and the security for delegates to
> assemble, conduct constitutional debates and craft political institutions
> to maintain stability and foster democracy for the longer term.
>
> (Roeder and Rothchild 2005: 2)

Second, these efforts also attempt to build arrangements for the longer
term, in the form of rewriting constitutions, even devising new
parameters for the 'state'. Most prominent among these have been the
'Taif Accord' in Lebanon (1989, http://www.al-bab.com/arab/docs/
lebanon/taif.htm) and the Good Friday Agreement in Northern Ireland
of 1998 (http://www.nio.gov.uk/the-agreement).

Third, they attempt to do so in a very public way to try and enhance
popular support for the democratization process, as well as to persuade those who have been demobilized that they have to change
their behaviour. DDR mechanisms are thus required for the peacebuilding process to be seen to be working. Not to do so will lead to
'stigmatizing' ex-combatants (McMullin 2013a).

The importance of, and potential for, DDR and SSR in using the tools
of development to try and transform conflicts is therefore huge. Their
aims are both on the individual and the collective level. On the
individual level the aim is the transformation of rebel movements into
peacetime security forces or other economic groups and the demobilization of huge numbers of young people who have often known of
nothing but war and killing in their short lives. This individual
demobilization has to provide extensive educational, social and
psychological help to young ex-combatants (Peters 2007). The group
from which they come then has to be transformed into a political
party that relies on the ballot box, not the gun, for its legitimacy. De
Zeeuw is succinct in his assessment of the likely result of this:
'Experiences from people directly involved in the transformation of

such movements show that the process is extremely complex and time-consuming and has a high risk of failure' (de Zeeuw 2008: 1). This seems to be particularly true of individuals who have spent a great deal of time in the field, often as children who then graduate to leadership roles in insurgent armies. One such example is the (Ugandan-based) Lord's Resistance Army (led by Joseph Kony) Deputy Commander Dominic Ongwen, who was kidnapped as child of ten and went on to become Kony's right-hand man. Ongwen is at the time of writing awaiting trial before the ICC in The Hague (http://www.bbc.co.uk/news/world-africa-30906019).

The issue is so important because the notion of criminal liability is said to be difficult to prove against a child, yet there are differing views about the age that a 'child' becomes an 'adult'. The question is often posed as to how 'innocent' such young adults are, how much they should be held responsible for their actions and to what extent they are 'victims' as well as 'perpetrators' (Shepler 2005). In addition, IGOs tend to reduce help to 'adults' once they come of age, so consigning them to a nebulous hinterland of stigmatization at an arbitrary age. The risks of recidivism are very strong (McMullin 2011, 2011a).

The results have been mixed. In Sierra Leone the difficulties of transforming the Revolutionary United Front (RUF), the rebel group led by Foday Sankoh, into a 'less violent political organization' failed until the UN peace mission effectively detained 'a large proportion of the movement's more politically articulate elements' (Richards and Vincent in de Zeeuw 2008: 91). RUF foot soldiers have not, on the whole, been retrained, except within the new regime's armed forces and police. Richards and Vincent conclude that '[u]nless the problems of an expanding and increasingly impoverished youth underclass are addressed, violent instability is likely to return to the forests and diamond districts of rural Sierra Leone' (Richards and Vincent in de Zeeuw 2008: 100). There is also a long-term legacy of mental instability among many of the former child soldiers in Sierra Leone, whose results will take many years to be fully appreciated (Betancourt et al. 2010).

On the collective level DDR and SSR aim to provide the basis for a lasting political settlement, through such processes as constitution writing (Caplan 2005: 29), which have occurred in many states in the past 35 years, nearly always internationally brokered by IGOs like the Commonwealth or think tanks like the United States Institute for Peace (Widner 2005). It is a complicated procedural process that often drags on for years during or, sometimes, after the conflict is mainly 'over', and aims 'to develop a sense of inclusion and trust (social capital)' in a process that will hopefully encourage peaceful dialogue, not conflict, through providing frameworks of cooperation where different sections of the population feel their interests are being acknowledged. This content of representation (after Pitkin 1967) has been found by some researchers to have 'no major effects on post-ratification levels of violence in some parts of the world, such as Europe, but do[es] make a difference in Africa, the Americas and the Pacific together' (Widner 2005: 516). This would seem to indicate that such efforts have some empirical evidence to back up their claims.

As Roeder and Rothchild admit, the underlying rationale for these actions is a belief that democracy is the essential tool that can calm the fervid spirits of war. But what if the erstwhile combatants have been divided by religious, ethnic or other problems that might lead them democratically to want to *divide from, not unite with*, each other? 'Majoritarian democracy can be a potential source of heightened interethnic conflict' (Roeder and Rothchild 2005: 5). So one solution that has been widely touted is 'power sharing', as was widely tried in the 1990s.

What (else) is missing from the democracy, DDR and reconstruction analyses?

The following issues will be discussed:

- Having a peace to keep?
- 'Spoiler violence' in post-conflict reconstruction
- The challenge of small arms proliferation
- How can former combatants be 'demobilized' and 'reintegrated'?
- The challenge of public health
- The challenge of the presence of outside actors.

Having a Peace to Keep?

The main 'official' issues identified in most DDR operations by the UN have been with

1) problems associated with the establishment and maintenance of a security environment early on, and 2) problems concerned with a lack of coordination efforts among the regional and international communities,

> the various groups involved in a peace mission, the peace mission
> itself, and the post-conflict reconstruction effort.
>   (UNIDIR 1996–8: 211, quoted by Gamba in Darby 2006: 54)

Democratization, DDR and reconstruction processes assume that the
war is 'over' or at least in its final throes. But is there not (again) an
'artificial distinction between armed conflict and post-conflict'? And
are the international organizations putting an excessive emphasis on
DDR policies as 'magic bullets' (Muggah 2005)? There are therefore
a number of grounds on which the above democracy and DDR
orthodoxy can be challenged. Some of them are generic and some
(even many) more are based on the empirical observation of case
studies.

We can also point to a more complex problem. As we have stressed,
the causes and results of wars are intimately linked (Blainey 1988).
But what if the DDR process itself actually makes people who are
already very upset even more so? Grievances from before a conflict
are carried into it and beyond it. Can the bringing of democratic
institutions persuade combatants to feel that they can now lay down
their weapons and all these hatreds will just evaporate? Keen's
analysis that 'greed' and 'grievance' are not sufficient to explain why
conflict breaks out, with which the authors broadly agree, means that
a very deep-rooted analysis has to be made about all actors' responsibility for the fighting that has taken/is taking place, not just the
militias (Keen 2007). Maybe, as with Sierra Leone, the elders pro-
voked violence by their insensitivity over a long period to the
demands of youth. Are they to be the new 'Members of Parliament'?
How does a democratic process address a perceived lack of 'respect'
agenda? How does the DDR process address a cycle of abuse, maybe
over many generations? How do they know they will not be further
'betrayed' (Keen 2007: Chapter 3)?

'Spoiler violence' in post-conflict reconstruction

In addition, in the aftermath of a number of civil wars and conflict in
developing countries, it has been found that the main occupation of
many of the youthful population of the country has ceased to be 'normal', but rather to revolve around membership of various militias,
groups that often indulge in 'freelance' work on the side. As Podder
writes (2012: 199), '[r]eintegration is accepted as the weakest link in

DDR efforts.' Partly this is due to the mental scars that the former combatants carry with them (as discussed above), partly to the difficulty of getting communities to re-engage with those who have previously done them harm. And partly it is due to former combatants not being so 'former': in Afghanistan, for example, former Taliban and other militia, or 'non state' armed forces, members took a long time to demobilize after the Bonn Accords of 2001, and often resumed their former activities, under cover of nightfall, for example (Giustozzi, in de Zeeuw 2008). This was in spite of there being a new constitution in 2004 and elections. On a more positive note, although youthful badly 'demobilized' soldiers can become the thugs of the new 'parties' who intimidate their opponents, they can also form the nucleus at pro-democratic party rallies.

One related obvious area (dealt with elsewhere in this volume), is that of 'post-conflict crime' and what is referred to as 'spoiler violence' by disaffected paramilitary groups or even by 'official groups'. Sometimes it is driven by a deliberate attempt to undermine peace accords, sometimes by more venal motivations, but the results are identical. The point is that such violence seems attendant on most, if not all, post-conflict situations. It has been defined as 'violence that deliberately attempts to undermine peacemaking processes and peace accords' (Mac Ginty 2006) and it clearly will have a major impact on attempts at post-conflict reconstruction. However, as John Darby has put it, 'although substantial research attention has been paid to the origin and dynamics of ethnic violence, to the first moves towards negotiations, and to spoiler violence, the threat to post-accord reconstruction is under-researched' (Darby 2006: 6).

Spoiler violence can be generated by unofficial actors (militias, criminal groups, etc.) who feel that the DDR process has failed them, as in Liberia in the 1990s, or by the state itself provoking violence, as in Zimbabwe with the taking of land from white farmers in clear violation of the 1979 Lancaster House Agreement, or the government of Rwanda in 1994 by provoking the genocide. Often this is because a conflict in its final or initial post-conflict stage is still one where groups contesting established governmental power are jockeying for position for both a negotiated outcome and on the 'battlefield' while the state is trying to do the same (Höglund and Zartman in Darby 2006). The Taliban was not 'defeated' in 2001 and has been trying to bomb its way back into Afghan political life; it certainly does not consider the war 'over'. Neither do the various parties destabilizing Iraq, whether they be disaffected ex-soldiers and officials of the Saddam Hussein regime, or other Islamic insurgents like ISIS *(Daesh)*. In other words, the parties and those attempting to 'manage' or 'resolve' the conflict do not operate on one sole path, but rather on parallel paths. Conflicts are not resolved in nice neat linear ways, but by a process of trial and error, more like a spiral than anything else (Lederach 1997).

It must also be said that other forms of 'spoiler violence' are also deeply ideologically satisfying for groups that feel they have not been given adequate recognition in the process of reintroducing structures of government. In the chaos brought about by the 2003 invasion of Iraq, for example, very little thought was given to what would be done about demobilized members of the Iraqi armed forces (see Box 5.2). It was certainly also not considered that foreign insurgents would join in the aftermath of the 'end' of the war pro-claimed in May 2003 by President Bush, again for ideological reasons. Insurgents in Iraq, like *Daesh*, have taken the concept of spoiler violence to a terrible new level and made a mockery of the idea that the 'liberal peace' can easily bring democracy to such divided societies. Indeed it could be argued that the only place where democracy is now flourishing in the area is Kurdistan, where a group that has been marginalized and oppressed since at least 1919 by three states (Iran, Iraq and Turkey) has finally found its voice, but through the barrel of a gun not through patient democratic peacebuilding (King 2013; Voller 2014).

We shall return in the conclusion to this chapter to what this might say about future prospects for such massive 'nationbuilding' projects backed by military force and with a clearly defined economic and political agenda.

The challenge of small arms proliferation

Many writers (Duffield 2001; Muggah 2005; Greene and Taylor 2012; Kaldor 2012; Batchelor and Kenkel 2013 are just a few exam-ples) have noted that recent conflicts leave countries awash with small arms. In Northern Ireland, where great efforts have been made to 'put weapons beyond use', that has not made them disappear for ever. Globalization has ensured that small arms get sucked into

Box 5.2

The problems of reconstruction and DDR in Iraq, 2003–15

The official 'end' of the war in Iraq in May 2003 was followed by a demobilization of nearly all the Iraqi armed forces and a purging of practically all of the former regime's officials, with an interim Coalition Provisional Authority that became an Iraqi government after elections in January 2005. The country has since collapsed into civil war with an estimated 200,000 civilians killed in intra-communal violence, an Al-Qaeda inspired insurgency, the actions of the international interventionary forces and by general criminality. Very little of the many billions of dollars disbursed in reconstruction funds has been used for that purpose, but rather has been lost to corruption or to the sheer cost of security. The CPA only spent a fraction of the $18.6 billion allocated for reconstruction by the US Congress for that purpose for example in 2003–4 alone (Diamond in Fukuyama 2006: 176). It has since emerged that all the preparation for the transition by the Future of Iraq Project within the State Department has been essentially ignored and the Pentagon has been allowed to dictate a failed post-war policy, the rule of law flouted by the occupying forces (epitomized by atrocities by US troops in Abu Ghraib prison) and the Shia and Sunni militias alike. The only relatively peaceful area of Iraq after 2003 is that of Kurdistan, which was effectively independently governed from the mid-1990s anyway. A 'surge' in US forces in 2007 led to a reduction in violence but the arrival of Islamic State (Daesh), and the spillover of the Syrian civil war has led to overt civil war between different factions and especially between Shia and Sunni hardliners. Tensions that already existed with Iraq's neighbours Iran (accused of aiding Shia violence by the US), Turkey (which has been attacking Kurdish irredentists in the North) and Saudi Arabia (accused of encouraging extreme Sunni ideology, if not of actually supporting Sunni insurgents) have grown into widespread terror and civil war displacing millions of people and killing and maiming many more. Iraq has become a battleground for regional and global differences, thus complicating the reconstruction efforts immeasurably.

Sources: Dobbins et al. (2007), Allawi (2007), Dodge (2013),, Duffield (2007), Dodge (2006), Ismael and Ismael (2005), Rogers (2013), Sky (2015).

conflicts far more easily than was the case in the Cold War, where at least the Superpowers acted as some sort of gatekeepers for violence and where states were propped up, not, as they now are, in a state of collapse. This is not to minimize the horrible effects of Superpower actions in many parts of the world, with the obvious after-effects of such actions still very visible in South East Asia, the Horn of Africa and elsewhere – we must be careful not to fall into Cold War nostalgia.

Plate 12 *'No weapons' sign at the entrance to a UN facility: disarmament, demobilization and reintegration are often among the thorniest of post-war problems.*

A publication of the Small Arms Survey group, based in Geneva (2015), showed that:

> violent deaths decreased globally; yet armed conflict grows more lethal. This edition estimates 508,000 violent deaths per year between 2007–12, down from 526,000 reported in 2011 for the period 2004–09. A larger proportion of these deaths, however, were directly related to conflict (70,000 deaths per year, up from 55,000).
>
> (http://www.smallarmssurvey.org/about-us/highlights/ highlights-2015/highlight-gbav-press.html)

In particular the report indicated that the vast majority of such deaths were in 18 countries and that 25 per cent of them in 2012 (the latest date for which there were accurate statistics) were attributable to Syria, Honduras and Venezuela.

But the evidence of the destabilization of many people's lives has particularly been evident in Africa where 'traditional violence' has been horribly incremented by the dissemination of huge quantities of small arms, often of former Soviet Bloc and Chinese manufacture. The worst current outbreak is in Northern Nigeria, Cameroon, Chad and Niger where the Islamic group Boko Haram has unleashed a war of terror against the population since 2002 (International Crisis Group 2014a). This in turn has destabilized whole societies and had devastating effects on the men, women and children involved, as well as on the stability of the states concerned (Mkutu 2008; Wepundi et al. 2012). So yet again we have the paradox of fewer actual 'wars' but increased insecurity encouraged by the after-effects of 'old wars' that stimulate the lower level of violence that is a feature of the 'new' ones.

In the case of Northern Ireland, Colin McInnes (2000) has identified a number of issues that have to be addressed when decommissioning weapons. They can act as shorthand for other cases (McInnes 2000; Fitz-Gerald and Mason 2005):

1. WHEN to decommission? Before or after the final political agreement? Should this be a prerequisite for negotiation?
2. LINKAGE to other issues? In Northern Ireland, those issues included release of prisoners. Problems included 'inequality' of concessions.
3. HOW to decommission – who provides data on weapons held; phasing of weapon surrender; to whom?
4. WHOSE weapons – those of 'insurgent groups'; 'Government' or 'occupying forces'?
5. WHAT weapons. Small arms, explosives…? N.B. 'the real 'weapon' is the skill and experience of those who made them'

(McInnes 2000: 89)

In general, what we can say is that disarmament is a phase that overlaps demobilization and reintegration. This is because the signing of a peace agreement often leads to large numbers of young men using their weapons in a freelance way – usually through preying on their own or other communities. In parts of Africa, where there are few if any effective border controls (or indeed clear borders), armed gangs of former combatants regularly cross into areas that have low policing, on missions of loot and rapine. This post-accord criminality is thus both a national and an international feature of efforts at DDR (Muggah 2005; Mac Ginty 2006).

How can ex-combatants be 'demobilized' and 'reintegrated'?

Proposed solutions to the problems outlined above have included the absorption of ex-combatants into existing police and military forces, even if DDR and SSR should clearly, and ideally, lead to far greater reintegration at other levels of society. It should therefore aim to 'help them develop alternative income-generating activities so they can provide for themselves and their families', though de Zeeuw comments that '[i]n many cases, the real socio-economic needs of the rank and file of former rebel organizations … are not addressed' (de Zeeuw 2008: 12–13). The World Bank and major Western governmental agencies like DFID have in particular urged 'best practice' on IGOs trying to implement such schemes. Surveys of related practice (as with Fitz-Gerald and Mason 2005) are as yet inconclusive, with evidence clearly tending to vary according to a number of factors. These can be said to include:

- The commitment of donor countries to the process.
- The conditions on the ground. So (we might say): in [country x] and [country y] there are reasons to be optimistic. In [country z] these conditions do not prevail, so [the following conclusions can be drawn]. We clearly have to be careful about over-generalization.
- The commitment to, and the ability to supervise 'Security Sector Reform' (SSR).

SSR is a particularly difficult area that has been increasingly studied and implemented in recent years, by both IGOs like the OSCE in Eastern Europe, as well as by civilian forces operated in conjunction with the UN more widely. Prominent among such organizations are the Global Facilitation Network for Security Sector Reform http://www.gsdrc.org/go/topic-guides/security-sector-reform run by the UK government's DFID (GFN-SSR 2007); the OECD (OECD 2007) and INSTRAW (UN International Research and Training Institute for the Advancement of Women; a UN agency that is particularly interested in the implications of SSR for gender relations).

GFN-SSR (2007) identifies four particular areas that need to be addressed in SSR:

1. Core Security Actors (police, gendarmerie, civil defence, militias, etc.)
2. Management and Oversight bodies

3. Justice and the Rule of Law
4. Non-statutory security forces ('liberation armies', private security organizations, etc.).

(GFN-SSR 2007)

It has to be said that in some cases the record of the international community in SSR has been very patchy. The Dayton Agreement of 1995 had important SSR elements in it, though Dayton has been severely criticized for leaving three separate armies in existence (Fitz-Gerald and Mason 2005: 13), a gap that has been addressed in Kosovo by setting up a police training school.

Some of the worst examples in terms of levels of violence can be found in the linked conflicts of Angola and the Democratic Republic of Congo (DRC) as well as in the Sudan, though there are many other candidates (a good source is the Geneva-based Small Arms Survey organization, http://www.smallarmssurvey.org/), which publishes an annual report. Angola was struggling to emerge from a civil war which dated back to independence from Portugal in 1975 (for a good overview see Cramer 2006: Chapter 4). The DRC has fallen into total, and heavily armed, chaos since the arrival of huge numbers of Rwandan *génocidaires* after the massacres of 1994 in Eastern Congo, some of whom have been supported by local and international actors with free supplies of cash and arms. The international community's efforts, through the United Nations Angola Verification Missions, UNAVEM I, II and III in the 1990s, to disarm the combatants in Angola was a humiliating failure, blamed by some on its inadequate budget and force levels (Gamba in Darby 2006: 61), but by others on its fundamental misreading of the nature of the civil war in which it was trying to intervene. Cramer (2006: 143) puts this down to the impossibility of rounding up the huge numbers of guns that have always flooded into Angola from the West since the seventeenth century and 'international linkages, political and economic' which are part of '500 years of violent conflict' (Cramer 2006:147). In effect, Cramer asserts that the spread of capitalism, through globalization, is the problem, and liberally minded IGOs can do little about this. We might also reiterate the point made earlier, that war is a profitable enterprise and that combatants have to be given very clear reasons why they would want to substitute that earning potential for the economic uncertainties of peace (Nordstrom 2004).

For even if it could be argued that such generalizations ignore the very real benefits that capitalism has brought for many previously poverty-struck developing countries, we also have to admit that the record of capitalism as a force for non-violent beneficial change is hard to show in much of Africa (see Box 5.3 on the DRC below), but that is demonstrably not the case in much of Asia, for example. In a globalizing world, the boundaries of where such spoiler violence can be found are unpredictable in the extreme. The same could be said of externally generated violence for more venal reasons in the DRC where freelance groups, multinational companies and different 'allied' or 'enemy' governments use the opportunity of the chaos to line their pockets with diamonds, coltan (a vital ingredient in mobile phones, see Pugh and Cooper 2004) and other raw materials. Without a viable state it seems clear that no war ever 'ends'.

Box 5.3

Disarmament, demobilization and reintegration in the Democratic Republic of Congo

The civil war in the DRC erupted after the genocide in Rwanda in 1994 forced many hundreds of thousands of Hutu extremists and refugees to flee to the DRC from Rwanda. The 'génocidaires', or 'Interhamwe', took their arms with them and set up enclaves from which they embarked on raids into neighbouring countries and preyed on the local populations, the national army being in no state to stop them. Other local militias formed and the area descended into anarchy, not helped by predatory capitalist entrepreneurs making the most of the anarchy to enrich themselves, as did the armies of Rwanda, Angola, Zimbabwe and Uganda, whether invited or not by the government in the capital, which had very little effective control over its eastern provinces. An 'All-Inclusive Accord' was signed between most of these warring parties in 2002 and agreed terms of reference for DDR. UN Peacekeepers were deployed to the eastern (Kivu and adjoining) areas and a semblance of national control re-established. The situation, however, remains very tentative, so although there have been serious attempts at DDR by the UN and other organizations, the situation in the east of the country is still chaotic, with local warlords regularly challenging the Blue Berets and DRC official forces, often successfully. Over four million people are considered to have been direct victims of the fighting and accompanying destruction. The record of DDR is mixed, with Kölln (2011: 18) reporting that of 2,400 child soldiers identified and processed by the Belgian Red Cross (one of many NGOs involved), only 553 had been demobilized and only 223 reunited with their families.

Sources: Boshoff, in Fitz-Gerald and Mason (2005), Pugh and Cooper (2004), Kölln (2011).

The challenge of public health

Developing countries generally have a very low ability to provide even basic levels of health care for their citizens, a problem exacerbated by conflict, which drives out both the central government as well as IGO and NGO provision. Therefore a clear potential beneficiary of the bringing of stable government to an area or region is that it will hopefully not only reduce levels of crime and violence but also help in the stabilization of the institutions of the state, which generally include educational, medical, and other social welfare facilities. The re-establishment of these institutions is often overlooked in writings on reconstruction, but by far the greater proportion of deaths and other suffering come from public health failures. Of the over four million victims so far of the civil war in the DRC, '[r]ather than battlefield deaths, most of these fatalities have been the result of disease and malnutrition as the state has failed to provide public health care or maintain sanitation and related infrastructure' (Mac Ginty and Williams 2005: 173). It is certain that in the calculation of 'war-related deaths' we have to take into account those caught in the crossfire of

Plate 13 *A UNICEF water tank in southern Lebanon: public health issues often pose a greater danger than direct violence.*

different groups, as well as the increase in morbidity due to people being displaced by the fighting, the spread of disease and the reduction of spending on health care by governments fighting for survival against insurgent groups (Muggah 2005: 240).

In some cases, failures of the health care system in developing countries can be the *cause* of men joining an insurgent group in the first place. One story we were told by a researcher in Sierra Leone was that a young man whose pregnant wife was refused treatment without a bribe at the (supposedly free) local hospital, as a result of which she died, used the first bullets from his new Kalashnikov to shoot up the maternity ward and its head doctor (Ahorsu 2007). It must also be understood that with the widespread use of rape as a weapon of war and the systematic abuse of women in war zones huge mental and physical health problems are caused that can take decades to heal, if they ever can (Nordstrom 1999; Brown 2012).

The challenge of the presence of outside actors

The success or failure of the liberal peace has at its heart the dilemma of *who* or *what* is to do the reconstructing? The assumption is that this will be done by the 'international community' working together in the United Nations and other IGOs, or subcontracting that to NGOs (Richmond and Carey 2005). This is what happened in the cases of Germany and Japan in the 1940s (Williams 2005a), and in those cases the local populations generally accepted the legitimacy of the actions taken. The same cannot necessarily be said of more contemporary examples.

The UN, IGOs and NGOs

The problem in many contemporary conflicts is that the international community acting through IGOs like the UN are rarely the only players. Their military muscle is generally weak or absent, and like NGOs they are reliant on peacekeeping forces to be able to operate (for the experiences of NGOs in East Timor and Sierra Leone, see Jackson in Richmond and Carey 2005). In some cases, NGOs and the United Nations are reliant on tacit or overt militia (even insurgent) support and are therefore both subject to often intolerable pressures to conform to what can be seen as 'alien' cultural practices by the local populations. Hence in Afghanistan during the period of rule by the Taliban, the UN and NGOs were in effect forced to comply with extreme *Sharia* law principles banning the unveiling, or even touching, of women. Even more seriously, in Iraq NGOs and the UN were forcibly put on notice that their presence would not be tolerated (Monshipouri in Richmond and Carey 2005). In this, albeit extreme, case the dissuasion was by the expedient of a large bomb which destroyed most of the UN HQ in Baghdad and killed its representative Sergio Vieira de Mello in 2004.

In less violent cases the presence of 'alternative' reconstruction groupings can directly challenge both the operations and the logic of liberal peace 'official' players. One example can be found in Lebanon where Hezbollah has its own 'reconstruction' arm, Jihad Al-Bina, and the Gulf states sponsor rival groupings, often to counter the ideological leanings of other groups. Such groups can rapidly marginalize the UN and conventional NGOs. The UN cannot, for example, distribute $12,000 in cash per household as Jihad Al-Bina did after the July 2006 war in Lebanon (Mac Ginty 2007: 458)

We would further suggest that there is an overall problem that has been better understood in the past few years. A growing trend towards the 'local' has become more marked in recent times as there has been an increase in the role taken by local communities and a corresponding loss of faith by these communities in traditional leaders or 'outsiders'. The lack of agency felt by many people who have been the victims of war is often due to the perception that the IGOs, capitalist institutions and firms and foreign governments are too self-interested to really care about local people. As has been said, '[l]ocal peacebuilding agency is not a new phenomenom... [but] it has become more prominent, and there is a greater willingness in some quarters to take it more seriously' (Mac Ginty and Richmond 2013: 773–4, 777). To put it briefly, local people know the circumstances of their suffering more deeply than could ever be the case with NGOs or IGOs from outside the conflict and they are increasingly demanding to be heard.

Conclusions

What can be done to improve matters? One depressing conclusion might be to say that all efforts at bringing about better outcomes in reconstruction and its allied strategies of DDR and SSR will always fail because of the necessarily violent nature of civil wars and endogenous statebuilding (Cramer 2006). A liberal peace interpretation might well be to say that there is a need for better institutional responses (for example Gamba, in Darby 2006, or Paris 2004). Chabal and Dalloz would point to an embedded enthnocentricity in such statements, an unwillingness to accept that we need to take cultural and historical context into account in condemning local practices or trying to impose 'a linear, when not a singular, form of modernization resulting in Westernization' (Chabal and Dalloz 2006: 9). A less extreme critical analytical position might be to say that we need to listen to local populations more (Mac Ginty 2006). A more 'structural' approach would point to the underlying economic problems that assail many developing countries. As we saw with the above discussion of Sierra Leone, DDR and SSR will only work if employment can be found for the vast underclass that is now a feature of so many developing countries. War, to put it bluntly, is profitable; peace is not (Nordstrom 2004).

More widely, it is clear that the misuse of the term 'reconstruction' in the early 2000s may well in the long run serve as the most hubristic of all misuses of a term, simultaneously devaluing a major historical success by misassociation and damaging hopes of any further use of the model in the future, a much greater potential problem as it is a basic ingredient of the policy toolbox of advocates of the 'liberal peace' (Williams 2007). We assert that the distinction that dates back to the American Civil War between accusations of 'carpet bagging' and the more noble term 'nationbuilding' used about Germany and Japan in 1945, still has some force in the latest attempt at 'reconstruction' in Afghanistan and Iraq in the 2000s (Williams 2006: 139–45).

We therefore hope to have shown that the political arguments for reconstruction lie on unstable historical foundations and have been further damaged by historical and recent experience, especially in Afghanistan and Iraq. In addition, economic arguments for intervention of all kinds have also been more fundamentally challenged from two key perspectives. Writers such as Cramer (2006) and Chabal and Dalloz (1999 and 2006) argue that the current debate on development in general and reconstruction in particular tend to downplay the importance of culture as a variable. This debate is arguably part of a much wider one, which suggests that there is no longer a coherent field of 'development studies' with 'competing schools of theory or paradigms' but that it has become subsumed within other disciplines (as with 'area studies') or hacked to pieces by those ('poststructuralists', 'post-modernists', 'post-marxists') who have under-mined all the 'positivist', for which read 'rational', bases for the study of development (Hoogvelt 1997: xi). One lasting impact of such epistemological uproar has been to make analysts (but not necessarily policymakers and politicians) focus more on what is specific to any reconstruction 'event' and less on what is generic, what might be termed a 'sociological turn'. We need to look at cases and their specificities, as well as to find 'one-size-fits-all' solutions.

It must also be remembered that some strategists such as Edward Luttwak (1999) and Stephen Van Evera (1999) think that the best and surest path to peace in most developing countries is to 'give war a chance'; in other words to not interfere in any way or form, but to let the naturally dominant party emerge though war. This might be put differently by asking: *when does a conflict 'end'?* As we have stressed, the concept of bringing about a 'sustainable' peace is at the heart of thinking about the end of a conflict. But, as a growing number of writers have pointed out, the 'ending' of a conflict is in itself a prob-lematic issue, and that therefore the concept of 'post-conflict develop-ment' and reconstruction is also. Both these perspectives are joined by one key commonality – the belief that war is the midwife of stable nation states, and it is the lack of such stability in various parts of the world that provides us with most of our problems. So we must look in every case at the underlying rationale for development and what makes for a successful nation in itself as a prelude to seeing what might go wrong with the process of 'state-' or even 'nation-building'.

The botched 'reconstruction' of Iraq will stand as the key case study that will inform all attempts at nationbuilding for the foreseeable future. It has been pointed out (Stiglitz 2008) that the cost of the war to the American taxpayer (with the attendant lost opportunity cost for American education, health care and other policy areas) has been, after five years of fighting, over $3 trillion. The cost to Iraqis in terms of their development and levels of life, has been arguably even worse since the first American intervention in 1991. One calculation has Iraqi per capita income dropping from $2279 in 1984 to $627 in 1991 and $450 by 1995, and levels like Madagascar and Rwanda immedi-ately before the 2003 war (Ismael and Ismael 2005: 613). Toby Dodge has pointed to the 'neo-conservative' belief that the United States would inherit a functioning Iraqi state, while it in fact collapsed with the invasion (Dodge 2006: 188). Charles Tripp goes further and suggests that the Iraqi state had in effect collapsed far earlier than

2003 and the looting of ministries was just the final act. But he also postulates that the Iraqi 'shadow state' organized along local lines of patronage has since 2003 been able to organize a 'headless' insurgency that has tied Coalition troops down and further degraded the lives of ordinary Iraqis to catastrophic levels. There is, he postulates, no democracy to defend (Tripp 2008, quoted in a lecture at St Andrews, 22 February 2008).

So what might be done to address these dilemmas? One way is to look back in the 'tool box' and ask what we have used in the past. Should the aim be to allow statebuilding in the developing world to progress independently of outside interference, if Münkler's point about the need for 'clinical' conditions is to be met? One idea from the Cold War period is that of 'self-reliance', a decoupling from globalization and a deliberate development of indigenous industry at the expense of manufactured imports, as was suggested in the 'New International Economic Order' of the 1970s in the United Nations Conference on Trade and Development (UNCTAD). But how would self-reliance stop the determined 'new' warlord intent on extracting his coltan or diamonds from a decrepit African state (Pugh and Cooper 2004)? On the other hand, there is much to be said for Galtung's point that the terms of trade that this exchange implies (Southern commodities for manufactured products from the West) means there is 'an enduring acceptance of a long-term inferior position in which it will be difficult to satisfy the basic needs of local people' (Galtung, paraphrased by Jeong 1999: 36).

In the next chapter it is hoped to take on these challenges and suggest how peace might conceivably be brought to developing countries by surveying approaches that attempt to examine the root, and especially the economic, causes and consequences of conflict to try and resolve deep-rooted processes both locally (from below) and from above through institutional action.

Summary

- The concept and practice of 'reconstruction' over many years has left it with a problematic image that cannot easily be repaired.
- Concepts and practices like 'state-' and 'nationbuilding' give rise to questions about the motivations and practices of those who try and implement them.

- One-size-fits-all approaches to reconstruction are unlikely to lead to optimal peacebuilding outcomes.
- Disarmament, Demobilization and Reintegration (DDR) as well as Security Sector Reform (SSR) lie at the heart of most, if not all, attempts at post-conflict reconstruction and are in themselves problematic.
- The international community needs to take more notice of the specific cultural and other needs and constraints of any particular situation before embarking on reconstruction attempts.

Discussion questions

- Why has reconstruction had such an uneven reputation for effectiveness in the past decade or so?
- Should the international community or powerful states attempt to 'reconstruct' economies and polities after wars?
- What is wrong, or right, with the proposition that democratizing a country will bring about a stable peace there?
- What are the main features of the practices known as 'DDR' and 'SSR'?
- How might the international community approach the question of reconstruction better in the future?

Further reading

Some of the best discussions of the problems and potential of reconstruction can be found in: R. Caplan (2005) *International Governance of War-Torn Territories: Rule and reconstruction*, Oxford: Oxford University Press; and A. Williams (2006) *Liberalism and War: The victors and the vanquished*, London: Routledge, Chapters 4 and 5. A more critical view is that of O. Richmond (2014) *Failed Statebuilding: Intervention, the state, and the dynamics of peace formation*, New Haven, NJ: Yale University Press. The best short explanation of why reconstruction attempts in Afghanistan and Iraq were conducted as they were can be found in J. Dobbins, J. G. McGinn et al. (2003) *America's Role in Nation-Building: From Germany to Iraq*, Santa Monica, CA: Rand Corporation, as well as J. Dobbins, S. G. Jones, K. Crane and B. Cole DeGrasse (2007) *The Beginner's Guide to Nation-Building*, Santa Monica, CA: Rand Corporation. One of the best critiques is T. Dodge (2013) *Iraq – From War to a New Authoritarianism*, London: Adelphi/Routledge, as well as his (2006) 'Iraq: the contradictions of exogenous state-building in historical perspective', *Third World Quarterly*, 27(1), pp. 187–200. The classic insider's view of what went wrong is R. Chandrasekaran (2007) *Imperial Life in the Emerald City: Inside Iraq's Green Zone*, London: Bloomsbury. This book was a key inspiration for Paul Greengrass's film *Green Zone* of 2010, starring Matt Damon. There are few defenders of what happened in practice, though one recent example of a sympathetic view is E. Sky (2015) *The Unravelling: High hopes and missed opportunities in Iraq*, London: Atlantic Books.

The best introductions to DDR and SSR are A. Özerdem (2008) *Becoming Civilian: Disarmament, demobilization and reintegration*, London: I.B. Tauris; A. Özerdem and S. Podder (eds) (2011) *Child Soldiers: From recruitment to reintegration*, Basingstoke: Palgrave; and J. R. McMullin (2013) *Ex-combatants and the Post-conflict State: Challenges of reintegration*, Houndmills: Palgrave Macmillan. See also A. M. Fitz-Gerald and H. Mason (eds) (2005) *From Conflict to Community: A combatant's return to citizenship*, Shrivenham: Global Facilitation Network for Security Sector Reform.

Useful websites

(All accessed as at 22 July 2015.) A number of other websites are cited in the chapter, so please also refer to those.

Data on current conflicts and wars can be found at Peace and Conflict website, University of Uppsala, http://www.pcr.uu.se/research/ucdp/datasets/. Details of some important peace accords can be found at 'Taif Accord' in Lebanon (1989, http://www.al-bab.com/arab/docs/lebanon/taif.htm) and the Good Friday Agreement in Northern Ireland of 1998 (http://webarchive.nationalarchives.gov.uk/20130128134220/http://www.nio.gov.uk/the-agreement). Some key reconstruction agencies can be found at the following sites: European Agency for Reconstruction (archived as of 2008) (http://ec.europa.eu/enlargement/archives/ear/home/default.htm); Department for International Development (UK) http://www.dfid.gov.uk/. On the UN's thinking on DDR see The United Nations Institute for Disarmament Research (UNIDIR, http://www.unidir.org). GFN-SSR can be found at http://www.ssrnetwork.net/.

References

Ahorsu, K. (2007) The political economy of post Cold War conflicts in Sub Sahara Africa: the natural resources factor. The case of the 1991 Sierra Leone civil war. PhD thesis, University of Kent.

Allawi, A. (2007) *The Occupation of Iraq: Winning the war, losing the peace*. New Haven, CT: Yale University Press.

Angell, N. (1910) *The Great Illusion: A study of the relation of military power in nations to their economic and social advantage*. London: William Heinemann.

Badie, B. (2014) *Le temps des Humiliés*. Paris: Odile Jacob; *Humiliation in International Relations: A pathology of contemporary international systems*. London: Bloomsbury (2016 forthcoming).

Batchelor, P. and Kenkel, K.M. (2013) *Controlling Small Arms: Consolidation, innovation and relevance in research and policy*. London: Routledge.

Beah, I. (2008) *A Long Way Gone: Memoirs of a boy soldier*. New York: Sarah Crichton Books.

Bell, D. (2007) *The Idea of Greater Britain: Empire and the future of world order, 1860–1900*. Princeton, NJ: Princeton University Press.

Bellamy, A. (2014). *The Responsibility to Protect: A defence*. Oxford: Oxford University Press.

Beretekeab R. (2000) *Eritrean Making of a Nation 1890–1991*. Uppsala: Uppsala University.

Beschloss, M. (2002) *The Conquerors: Roosevelt, Truman and the destruction of Hitler's Germany, 1941–1945*. New York: Simon & Schuster.

Betancourt, T.S. et al. (2008) High hopes, grim reality: reintegration and the education of former child soldiers in Sierra Leone. *Comparative Education Review* 52(4): 565–87.

Betancourt, T.S. et al. (2010) Sierra Leone's former child soldiers: a follow-up study of psychosocial adjustment and community reintegration. *Child Development* 81(4), July–August: 1077–95.

Blainey, G. (1988) *The Causes of War*, 3rd edn. New York: Free Press.

Boutros-Ghali, B. (1992) *An Agenda for Peace, Preventive diplomacy, peacemaking and peace-keeping*. New York: United Nations.

Brown, C. (2012) Rape as a weapon of war in the Democratic Republic of the Congo. *Torture* 22(1): 24–37.

Caplan, R. (2005) *International Governance of War-Torn Territories: Rule and reconstruction*. Oxford: Oxford University Press.

Carnegie Commission on Preventing Deadly Conflict (1997) *Preventing Deadly Conflict*. New York: Carnegie Corporation of New York.

Chabal, P. and Daloz, J.-P. (1999) *Africa Works: Disorder as a political instrument*. Oxford: James Currey.

Chabal, P. and Daloz, J.-P. (2006) *Culture Troubles: Politics and the interpretation of meaning*. London: Hurst.

Chandler, D. (2000) *Faking Democracy after Dayton*. 2nd edn. London: Pluto.

Collier, P. (1999) *Doing Well out of War*. Washington, DC: World Bank.

Cramer, C. (2006) *Civil War is not a Stupid Thing: Accounting for violence in developing countries*. London: Hurst.

Crocker, C.A., Hamson, F.O., Aall, P. (eds) (1996) *Managing Global Chaos: Sources of and responses to international conflict*. Washington, DC: USIP.

Crocker, C.A., Hampson, F.O., Aall, P. (eds) (2007) *Leashing the Dogs of War: Conflict management in a divided world*. Washington, DC: United States Institute of Peace Press.

Darby, J. (ed.) (2006) *Violence and Reconstruction*. Notre Dame, IN: University of Notre Dame Press.

De Goede, M. (2015) *Consuming Democracy: Local agencies and liberal peace in the Democratic Republic of Congo*. Leiden: Africa Studies Centre.

De Zeeuw, J. (ed.) (2008) *From Soldiers to Politicians: Transforming rebel movements after civil war*. Boulder, CO and London: Lynne Rienner.

Dobbins, J., Jones, S.G., Crane, K., Cole DeGrasse, B. (2007) *The Beginner's Guide to Nation-Building*. Santa Monica, CA: Rand Corporation.

Dobbins, J., McGinn J.G. et al. (2003) *America's Role in Nation-Building: From Germany to Iraq*. Santa Monica, CA: Rand Corporation.

Dodge, T. (2006) Iraq: the contradictions of exogenous state-building in historical perspective. *Third World Quarterly* 27(1): 187–200.

Dodge, T. (2013) *Iraq – From War to a New Authoritarianism*. London: Adelphi/Routledge.

Duffield, M. (2001) *Global Governance and the New Wars: The merging of development and security*. London: Zed.

Duffield, M. (2007) *Development, Security and Unending War: Governing the world of peoples*. Cambridge: Polity.

Durch, W., Holt, V., Earle C., Shanahan, M. (2003) *The Brahimi Report and the Future of UN Peace Operations*. Washington, DC: Henry L. Stimson Center.

ECLAC (United Nations Economic Commission for Latin America and the Caribbean) (2002) *Situación y Perspectivas 2002. Estudio Económico de América Latina y el Caribe, 2001–2002*, 1 August. Santiago: ECLAC. [Quoted in Augé, C. (2002), *Le Monde Diplomatique* [English edition], September.

Etzioni, A. (2015) COIN: a study of strategic illusion. *Small Wars & Insurgencies* (26)3: 345–76.

Fitz-Gerald, A.M. and Mason, H. (eds) (2005) *From Conflict to Community: A combatant's return to citizenship*. Shrivenham: Global Facilitation Network for Security Sector Reform.

Foner, E. (1989) *Reconstruction: America's unfinished revolution, 1863–1877*. New York: HarperCollins.

Fukuyama, F. (ed.) (2006) *Nation-Building: Beyond Afghanistan and Iraq*. Baltimore, MD: Johns Hopkins University Press.

GFN-SSR (2007) *A Beginner's Guide to Security Sector Reform*. Birmingham: GFN-SSR.

Ginifer, J. with input from Oliver, K. (2004) *Evaluation of the Conflict Prevention Pools: Country/Regional Case Study 3, Sierra Leone*. Bradford University, Channel Research Ltd, PARC & Associated Consultants.

Greene, O. and Taylor, N. (2012) *Small Arms, Crime and Conflict*. Abingdon: Routledge.

Hammond, P. (2007) *Framing Post-Cold War Conflicts: The media and international intervention*. Manchester: Manchester University Press.

Hehir, A. (2012) *The Responsibility to Protect: Rhetoric, reality and the future of humanitarian intervention*. London: Palgrave Macmillan.

Hoffman, D. (2011) *The War Machines: Young men and violence in Sierra Leone and Liberia (the cultures and practice of violence)*. Durham, NC: Duke University Press.

Hogan, M. (1987) *The Marshall Plan: America, Britain and the reconstruction of Western Europe, 1947–1952*. Cambridge: Cambridge University Press.

Hoogvelt, A. (1997) *Globalization and the Post-Colonial World*. London: Macmillan.

International Crisis Group Africa (2014) Curbing Violence in Nigeria (II): The Boko Haram Insurgency. Report 216, April.

Ismael, T.Y. and Ismael, J.S. (2005) Whither Iraq? Beyond Saddam, sanctions and occupation. *Third World Quarterly* 26(4–5): 609–29.

Jackson, M.G. (2005) A Necessary Collaboration: NGOs, Peacekeepers and Credible Military Force. The Case of Sierra Leone and East Timor. In Richmond, O. and Carey, H. (eds) *Subcontracting Peace: The challenges of NGO peacebuilding*. Aldershot: Ashgate, pp. 109–18.

Jacoby, T. (2007) Hegemony, modernisation and post-war reconstruction. *Global Society: Journal of Interdisciplinary International Relations* 21(4): 521–37.

Japan, Ministry of Foreign Affairs (MOFA) (2007) Japan's efforts on peacebuilding: Towards consolidation of peace and nation-building. Available at: http://www.mofa.go.jp/policy/un/pko/effort0704.pdf (accessed on 19 July 2015).

Jeong, H.-W. (1999) *Conflict Resolution: Dynamics, Process and Structure (Studies in Peace and Conflict Research)*. Farnham: Ashgate.

Jeong, H.-W. (2005) *Peacebuilding in Post-Conflict Societies: Strategy and process*. Boulder, CO and London: Lynne Rienner.

Kaldor, M. (2012) *New and Old Wars*. 3rd edn. Cambridge: Polity.

Keen, D. (2007) *Complex Emergencies*. Cambridge: Polity.

Keranen, O. (2014) Building states and identities in post-conflict states: symbolic practices in post-Dayton Bosnia. *Civil Wars* 16(4): 127–46.

Killick, J. (1997) *The United States and European Reconstruction, 1945–1960*. Edinburgh: Edinburgh University Press.

King, D. (2013) *Kurdistan on the Global Stage: Kinship, land, and community in Iraq*. New York: Rutgers University Press.

Kölln, A. (2011) *DDR in the Democratic Republic of Congo*. London: Peace Direct.

Lederach, J.P. (1997) *Building Peace: Sustainable reconciliation in divided societies*. Washington, DC: United States Institute of Peace Press.

Ledwidge, F. (2011) *Losing Small Wars: British military failure in Iraq and Afghanistan*. New Haven, CT: Yale University Press.

Lemarchand, R. (2009) *The Dynamics of Violence in Central Africa*. Philadelphia, PA: University of Pennsylvania Press.

Lewis, J. (2007) Nasty, brutish and in shorts? British colonial rule, violence and the historians of Mau Mau. *Round Table: The Commonwealth Journal of International Affairs* 96(389): 201–23.

Loewenberg, S. and Norby Bonde, B. (eds) (2007) *Media in Conflict Prevention and Peacebuilding Strategies*. Bonn: DW Media Services.

Lowe, K. (2013) *Savage Continent: Europe in the aftermath of World War II*. London: Picador.

Luttwak, E. (1999) Give war a chance. *Foreign Affairs* 78(4).

Mac Ginty, R. (2006) *No War, No Peace: The rejuvenation of stalled peace processes and peace accords*. London: Palgrave.

Mac Ginty, R. (2007) Reconstructing post-war Lebanon: a challenge to the liberal peace. *Conflict, Security and Development* 7(3): 457–82.

Mac Ginty, R. and Richmond, O. (2013) The local turn in peace building: a critical agenda for peace. *Third World Quarterly* 34(5): 763–83.

Mac Ginty, R. and Williams, A. (2005) Editorial: the Commonwealth and the inheritance of armed conflict. *Round Table: The Commonwealth Journal of International Affairs* 94(379): 173–5.

McInnes, C. (2000) A Farewell to Arms? Decommissioning and the Peace Process. In Cox, M., Guelke, A. and Stephens, F. (eds) *A Farewell to Arms? From 'long war' to long peace in Northern Ireland.* Manchester: Manchester University Press, pp. 78–92.

McMullin, J.R. (2011) Reintegrating young combatants: Do child-centred approaches leave children – and adults – behind? *Third World Quarterly* 32(4): 743–64.

McMullin, J.R. (2011a) Exclusion or Reintegration? Child Soldiers in Angola. In Özerdem A. and Podder S. (eds) *Child Soldiers: From recruitment to reintegration.* Basingstoke: Palgrave, pp. 359–87.

McMullin J.R. (2013) *Ex-combatants and the Post-conflict State: Challenges of reintegration.* Houndmills: Palgrave Macmillan.

Meredith, M. (2011) *The State of Africa: A history of the continent since Independence.* New York: Simon & Schuster.

Mkutu, K. (2008) Disarmament in Karamoja, Northern Uganda: is this a solution for localised violent inter and intra-communal conflict? *Round Table: The Commonwealth Journal of International Affairs* 97(394): 99–120.

Muggah, R. (2005) No magic bullet: a critical perspective on disarmament, demobilization and reintegration. *Round Table: The Commonwealth Journal of International Affairs* 94(379).

Münkler, H. (2004) *The New Wars.* Cambridge: Polity.

Münkler, H. (2005) *The New Wars.* Cambridge: Polity.

Nettlefield, L (2010) *Courting Democracy in Bosnia and Herzegovina: The Hague Tribunal's impact in a postwar state.* Cambridge: Cambridge University Press.

Nordstrom, C. (1999) Girls and War Zones: Troubling Questions. In Indra, D. (ed.) *Engendering Forced Migration: Theory and practice.* New York: Berghahn, pp. 63–82.

Nordstrom, C. (2004) *Shadows of War: Violence, power, and international profiteering in the twenty-first century.* Berkeley, CA: University of California Press.

OECD (2007) *Handbook on 'Security Sector Reform'.* Paris: OECD.

Özerdem, A. (2008) *Becoming Civilian: Disarmament, demobilization and reintegration.* London: I.B. Tauris.

Özerdem, A and Podder, S. (eds) (2011) *Child Soldiers: From recruitment to reintegration.* Basingstoke: Palgrave.

Paris, R. (2004) *At War's End: Building peace after civil conflict.* Cambridge: Cambridge University Press.

Peters, K. (2007) Reintegration Support for Young Ex-Combatants: A Right or a Privilege? *International Migration,* 45(5): 35–59.

Pitkin, H.F. (1967) *The Concept of Representation.* Berkeley, CA: University of California Press.

Plamenatz, J. (1960) *On Alien Rule and Self-Government.* London: Longmans.

Podder, S. (2012) From recruitment to reintegration: communities and ex-combatants in post-conflict Liberia. *International Peacekeeping* 19(2): 186–202.

Pugh, M. and Cooper, N. with Goodhand, J. (2004) *War Economies in a Regional Context: Challenges of transformation.* Boulder, CO: Lynne Rienner.

Richmond, O. and Carey, H. (eds) (2005) *Subcontracting Peace: The challenges of NGO peacebuilding.* Aldershot: Ashgate.

Roeder, P.G. and Rothchild, D. (eds) (2005) *Sustainable Peace: Power and democracy after civil war*. Ithaca, NY and London: Cornell University Press.

Rogers, P. (2013) Lost cause: consequences and implications of the war on terror. *Critical Studies on Terrorism* 6(1): 1–17.

Shepler, S. (2005) The rites of the child: global discourses of youth and reintegrating child soldiers in Sierra Leone. *Journal of Human Rights* 4: 197–211.

Sisk, T.D. (2013) Power-sharing in civil war: puzzles of peacemaking and peacebuilding. *Civil Wars* 15 Supplement 1: 7–20.

Sky, E. (2015) *The Unravelling: High hopes and missed opportunities in Iraq*. London: Atlantic Books.

Slomp, G. (2008) On Sovereignty. In Salmon, T. and Imber, M. (eds) *Issues in International Relations*. London: Routledge.

Small Arms Survey (2105) *Global Burden of Armed Violence 2015: Every body counts*. Geneva: Small Arms Survey.

Stiglitz, J. (2003) *Globalization and its Discontents*. New York: Norton.

Stiglitz, J. (2008) *The Three Trillion Dollar War: The True Cost of the Iraq Conflict*. New York: W.W. Norton.

Thomas, M. (2014) *Fight or Flight: Britain, France and their roads from empire*. Oxford: Oxford University Press.

Toal, G. and Dahlman, C.T. (2011) *Bosnia Remade: Ethnic cleansing and its reversal*. Oxford: Oxford University Press.

Tripp, C. The Local in Iraq. *Le Monde*. Available at: https://mondediplo.com/2008/01/02iraq (accessed on 9 December 2015).

United Nations Development Programme (2015) Post-2015 Sustainable Development Agenda. Available at: http://www.undp.org/content/undp/en/home/mdgoverview/post-2015-development-agenda/ (accessed on 19 July 2015).

United Nations Human Rights Commission (2015) *Report of the Commission of Inquiry on Human Rights in Eritrea*, A/HRC/29/42. New York: UNHRC.

Van Evera, S. (1999) *Causes of War: Power and the roots of conflict*. Cornell, NY: Cornell University Press.

Voller, Y. (2014) *The Kurdish Liberation Movement in Iraq: From insurgency to statehood*. London: Routledge.

Wepundi, M., Nthiga, E., Kabuu E., Murray, R., Alvazzi del Frate, A. (2012) *Availability of Small Arms and Perceptions of Security in Kenya: An assessment*. Geneva: Small Arms Survey.

Werbner, R. and Ranger, T. (eds) (1996) *Postcolonial Identities in Africa*. London: Zed.

Wiafe-Amoako, F. and Mazrui, A. (2014) *Human Security and Sierra Leone's Post-Conflict Development*. Lanham, MD: Lexington.

Widner, J. (2005) Constitution writing and conflict resolution. *Round Table: The Commonwealth Journal of International Affairs* 94(381): 503–18.

Williams, A. (2005) 'Reconstruction' before the Marshall Plan. *Review of International Studies* 31: 541–58.

Williams, A. (2006) *Liberalism and War: The victors and the vanquished*. London: Routledge.

Williams, A. (2007) Reconstruction: the bringing of peace and plenty or occult imperialism? *Global Society: Journal of Interdisciplinary International Relations* 21(4): 539–51.

Winn, N. (ed.) (2004) *Neo-Medievalism and Civil Wars*. London: Frank Cass.

6 Development, aid and violent conflict

Introduction: how are aid and conflict prevention linked concepts?

Before the end of the Cold War it was widely, but not universally, believed by economists and conflict researchers alike that 'aid' would help developing countries to overcome their 'development' and 'conflict' problems in the broad senses of these terms. Even today some standard texts on conflict have very little mention of aid in their indexes. But since the end of the Cold War that omission makes increasingly less sense. The characteristics of the wars that we have identified have now put humanitarian agencies in the front line, both as distributors of conventional aid, such as foodstuffs, but also as potential participants, along with conventional government agencies and military forces, and paramilitary actors in rebuilding 'failed states' (Ghani and Lockhart 2008). The actors that try to help alleviate the suffering of those increasingly caught up in current wars are collectively the 'humanitarians' in Hoffmann and Weiss's term. 'External assistance' has become vitally important for governments and IGOs, and it has also become big business for firms who disburse or build infrastructure (Boyce and O'Donnell 2007). One important question that all who now study the conflict/development nexus are asking is whether they can provide any kind of solution to the evident suffering of developing countries' populations, or whether they are indeed part of the problem?

The most obvious victims of wars in developing countries are civilians, but sometimes so too are those who attempt to help them. Hoffmann and Weiss point out that not only are 'the victims of war... the actors' intended targets', but also '[h]umanitarian organizations have reacted to the new wars but have not adapted' (Hoffmann and Weiss 2006: xvi–xvii). Other writers go further and see the aid agencies as part of a new pattern of global governance by the North whereby the populations of the South are made to feel their 'exclusion' in new ways and to enforce the ideological precepts of the liberal peace, in effect as agents of a new 'imperialism'. According to this view, underdevelopment has come to be seen as a security threat, not just a humanitarian disgrace, and consequently all those who try to correct this insecurity are in effect complicit in such a logic (Duffield 2001 and 2007; Easterly 2006). Equally problematic, these and other observers see aid as actually *fuelling* wars, keeping them going. Carolyn Nordstrom reports a conversation she had in Angola with a local youth during the civil war – 'Peace? Forget it, there's too much money being made here' (Nordstrom 2004: 191) – as one example of such feelings. In short, aid since about 2001 has had a heavy security component, it has been 'securitized' (Duffield 2007) and is no longer seen as solely the charitable action it once was, and still is for many.

There is an increasing awareness that an unfocused and indiscriminate distribution of aid can actually harm the statebuilding process that is the stated prime objective of many IGOs and NGOs (OECD 2010). Since the withdrawal of interventionary forces in Afghanistan and Iraq and the upsurge of fighting in Libya and Syria, all of which have happened in the period since 2011, there is also the problem that aid has largely stopped, or been drastically reduced leaving societies made vulnerable by war even more exposed. Aid agencies face a crisis of access. They cannot access populations in need to assess the situation, nor deliver food or medicine.

We have therefore chosen to extract the idea of 'aid' from our wider discussion of conflict in developing countries, because we believe that, like with issues of gender, health and other 'personal' issues, the problematic of aid's motivations and delivery gives us a powerful series of perspectives on what is right or wrong with current approaches to conflict and development. So aid provides another link to understanding the dynamics of conflict. It is undeniable that aid is a key part of the political economy of wars and as the nature of wars changes, so does that of aid, as well as comment about it.

The purpose and history of aid

The *purpose* of aid has been defined, first, as an:

> international social contract… a broad understanding amongst developed countries that, in order for the world to be, or to be seen to be, a

moderately equitable place, or at least to alleviate some of the worst suffering, there needs to be some form of international assistance.

(Hunt 2004)

Janet Hunt points out that, as well as this laudable aim, '[m]any donors provide aid not only for humanitarian reasons, but to enhance their own economic and political interests, through encouraging their own exports, or shaping the economic policies or political persuasion of recipient countries (Hunt 2004: 67). It could be seen as yet another weapon in the arsenal of 'economic statecraft', to use the famous phrase coined by David Baldwin (Baldwin 1985).

Aid is usually seen as being part of a wider attempt to arm societies against negative influences – to develop their economies and in particular to help them pursue liberal democratic peace strategies. Since 1990, aid has been seen as using the 'anticipated "peace dividend" to repair the ravages of the superpower competition in many war-torn and conflict-prone societies' (Forman and Patrick 2000: 2) and as an integral part of the liberal peace tool kit (Marriage 2008: 5–6).

Aid in historical perspective

In one of its first formulations as a result of the Marshall Plan (the full title of which was the 'European Recovery Program', initiated by the US Foreign Assistance Act of 1948), Marshall 'aid' was open to the same criticisms as those levelled against 'reconstruction' more recently. It was seen as politically motivated, had 'strings' attached which tied the recipient into an economic and political system that they did not necessarily want and it was often only a sticking plaster on a big wound. Nonetheless, until the 1970s 'aid' was seen by non-Marxists as a largely unproblematic extension of the idea of 'charity', helping out those less fortunate than oneself. As a child, I was told that any food I did not eat would be sent by parcel to help the 'poor children of Africa'. But in the 1970s that critique got more intellectually problematic; it was now asserted that aid gave rise to 'dependency' and that was as bad a thing in LDCs as it was on the streets of London. Peter Bauer revived the idea that trade, not aid, was the key to development. He saw the positive examples of countries like Singapore and other 'Newly Industrialized Countries' (NICs) forging ahead by opening up their economies and the progressive sclerosis affecting the socialist developing countries and those who still put their faith in handouts from the rich countries (Bauer 1991). This view was, and is, unpopular in some quarters – the *Guardian* described Bauer as 'the shrillest Thatcherite spokesman against development aid for the third world' (*Guardian* 2002).

Both Superpowers used the idea of 'aid' in their own ways to buy influence in the newly emerging developing world. The USSR set up 'radial' trade and development agreements (so called as they radiated from the hub of the Soviet Union), of which one of the most notorious was the provision of cultural, industrial and security assistance (including nuclear arms) to Communist Cuba, in return for Cuba providing much of the USSR's needs for sugar, tobacco and a security base in the Caribbean. The USA provided huge amounts of Marshall Aid to Western Europe and followed that up with technical assistance, soft loans, etc. as well as military 'advice' (Seitz 2012). This activity was predicated on the American success in rebuilding its own economy after the Great Depression. It also fell away after the debacle in Vietnam in 1975 and the rise of the new economic orthodoxy of free markets epitomized by President Ronald Reagan and British Prime Minister Margaret Thatcher, as will be explored further below (Ekbladh and Sutton in Fukuyama 2005).

The economics and politics of aid: the case for intervention in developing countries

In the context of a book like this, a first set of questions about the case for intervention has to take into account both the economic and the political. These have tended to be put into separate disciplinary boxes, but such an approach can be seen as one-sided and prone to leading us into many misapprehensions and misconceptions.

The economic case for intervention

British Labour Party politician Jack Straw summed up why there is a widespread belief that poverty and conflict overlap:

> Look at where there are people living in poverty, on less than $3 a day and where there is conflict. The overlap is an exact fit. And look at where people live in prosperity and lack of conflict. Again the fit is exact.
> (Straw, BBC 10 February 2007)

This has been progressively explored, especially since 2000, especially in the 'greed and grievance' debate developed by economist Paul Collier, who suggests that war is either fuelled by a sense of 'grievance' for real or perceived social, economic or political injustice or by a simple desire of one or more group within a conflict-torn society to benefit from the turmoil to amass the most resources possible. A classic example of this latter impulse was the accumulation of 'blood' diamonds by different groups in West Africa (Zoellner 2006; Bieri 2010).

Collier's contribution to the debate on aid includes, most notably, his book *The Bottom Billion*, which posited that aid had to be an important part of using 'carrots' and 'sticks' in encouraging developing countries towards better governance and development, the royal roads to better security (Collier 2007). As Collier put it in a talk at Monterey in May 2008, in his view the mantra 'Aid–Trade–Security and Governance' (the 'ATSG Mantra') has to be followed in dealing with societies that have been stuck at the 'bottom', by which he means at the bottom of the economic and political pile, for 40 years (Collier 2008).

Collier's 'ATSG Mantra' was reinforced by major liberal economists like Jeffrey Sachs, whose equally uncompromising *The End of Poverty: Economic possibilities for our time* (Sachs 2005) also picked up on the 'billion' souls to be saved from themselves. For both Collier and Sachs the explicit model being emulated is that of the United States' actions in the 1940s. Europe was saved by aid, *ergo* Africa can be too. Sachs's work is best summarized in his own words: 'War can … erupt as a result of the collapse of an impoverished society, one suffering the scourges of drought, hunger, lack of jobs, and lack of hope. Ending poverty is therefore a basic matter of our own security' (Sachs 2007).

In much of the writing on development there is an implicit or explicit reference to the economic causes of conflict that has made it necessary, or to the potential healing power of economic action. Initially, all of these explanations ask what is it that makes for, in David Landes's (1998) term, the 'wealth and poverty of nations'? His essential point is incontestable – states and regions are differently endowed with economic assets and resources by virtue of nature's distribution of wealth. Landes points to the now somewhat discredited, but still valid, axiom that 'geography, especially climate, influence[es] human development' (Landes 1998: 3). It has been crudely suggested that hot climates can lead to less ardour in the workplace and the 'productivity of labor in tropical countries was reduced accordingly', a form of nineteenth-century environmental determinism that many would find distasteful. Disease, high morbidity, especially among infants, poor water supplies have added to this obvious burden. (Landes 1998: Chapter 1 'Nature's Inequalities'. See also Chapter 1 of this volume for more discussion on the human side of underdevelopment.)

Classical economists (of whom Landes is one) give us to believe that states emerge from a natural process of a locally and internationally logical process of optimal allocation of resources, including land, capital and labour. Hence 'mature' economies have tended to go though a process of development. What is wrong with this? First, the 'mature' economies all flourished in war and peace under a strong protective shield during their early development, whereas in contemporary times most developing countries have been urged to compete in international markets without such a benefit. Second (and more importantly), societies do not just get to choose their economic policies in a rational way. Some societies, like Switzerland, have managed to survive and prosper in spite of having virtually no economic advantages – mostly low-grade and mountainous land areas, endowed with poor natural resources, afflicted by linguistic and religious difficulties and surrounded by powerful and unfriendly neighbours, all factors that have on occasion engendered serious civil strife. Yet there are others, like Nigeria and Russia, that are endowed with enormous natural resource bases and educated populations that seem to be in a state of perpetual strife. Some slightly 'populist' economists like Jared Diamond point to societies that implode through ecological hubris and even 'passivity' faced with overwhelming challenges. Size is on occasion a benefit and in others a problem (Diamond 2005).

Of course, globalization might be said to have exacerbated both the differences and the potential benefits and disadvantages of such initial economic profiles. In the first wave of globalization in the nineteenth century, this became concretized in an imperial relationship often based on economic exploitation. In its earlier manifestations, in the nineteenth century, globalization has been identified by historians as being categorized as 'archaic, proto, modern and post-colonial' (Hopkins 2002).

In the latest wave of globalization since the Second World War, the economic relationship has arguably not changed much, but has been replaced by a 'centre–periphery' relationship often based on the 'centre' being well endowed with capital and technology and the periphery with a wealth of primary resources and cheap labour. This is not a problem, or indeed a surprise to classical economists, but of course it does create tensions that often come out in the apportioning of blame for both underdevelopment and the ensuing conflict. Scholars have asked repeatedly if globalization therefore leads to 'convergence or divergence' of societies and economies (Hülsemeyer 2003).

Is it all about 'economic readjustment'?

The debate about whether the pursuit of market reform helps or hinders development in developing countries hinges on whether the market alleviates one of the main causes of conflict, widely identified as being *poverty*. The encouragement of trade plays a significant role in liberal thinking about comparative advantage (Adam Smith 1776), as well as that trade links encourage peace (Mill, Bright, etc.). As Jacoby has succinctly put it: '[l]iving freely is thus trading freely' (Jacoby 2007: 524). So what is, potentially at least, wrong with the assumption that if we get the economics of a developing country emerging from war right, we will then get the politics right too?

To put it in a nutshell, if the recipients (or 'beneficiaries') of market reforms are of the clear opinion that they will not be better off as a result of them, they will tend to lump together their sense of grievance, from whatever source that comes, with the message that their main persecution comes from a capitalist, globalized world system. If they are then told that they must comply with a policy of 'reconstruction', they will then reject the whole package. Therefore, as with the *causes, escalation and maintenance* of conflict (Pugh and Cooper 2004) outlined in Chapter 1, so with the attempts to *end* it. We will explore the reasoning of why intervention might end up making conflict worse rather than better with a more in-depth examination of the economic arguments below.

In contrast to the 'Keynesian' interventionist, import-substitution, economic policies pursued by Western states and the IGOs alike until about 1980, a policy based on much more unfettered free market principles prevailed after that date. This is often referred to as 'Reaganomics' (after President Ronald Reagan, 1980–8), and was also espoused by UK Prime Minister Margaret Thatcher (who served

Plate 14 *Repairing war-damaged housing in Bosnia: who should pay for this, and what role should private enterprise play?*

between 1979 and 1990). This way of thinking had profound effects on domestic economic policies across the West. It also had a huge impact on the thinking of international institutions like the World Bank and the IMF in the 1980s and led to the promotion of export-led growth by developing countries as opposed to the 'import-substitution' models previously fashionable. The result was that the 'self-reliance' models of the 1960s and 1970s popularized by economists like Raul Prebsich and institutions like the United Nations Conference on Trade and Development (UNCTAD) were jettisoned in favour of policies to integrate developing countries into the international trading system, in a conscious imitation of the seeming successes of the 'newly industrialized countries' (NICs) like Singapore and South Korea. This was the beginning of what we would now call 'neo-liberalism'. If the NICs could grow through exports and inward investment, why could not Ghana, India and Tanzania (for example) do the same (Toye 2014)? The answer was of course that the economic advantages of small entrepôt states (Hong Kong and Singapore) and ones backed up by the might of the United States (South Korea) were quite different from those of poor, conflict-ridden states in Africa. But economic ideology and fashion did their work and the result was that aid (usually in the form of World Bank loans) was used to fund 'white elephant' investment projects across the Third World in the 1980s. Dams came to symbolize both the hope of this new vision of aid and also its failings (Usher 1997). Dams were often disastrously bad for the environment, drained foreign exchange reserves of the recipient states, encouraged corruption and often did not even produce much useful electricity, as with the British-sponsored Pergau Dam in Malaysia (Lankester 2012). The Aswan Dam in Egypt is one obvious example of both drawbacks, but there are many more.

At much the same time, and for many of the same ideological reasons, forceful economic arguments against direct aid were made by writers like Bauer (Bauer 1991), who claimed it led to a form of welfare dependency similar to that experienced by benefit-dependent dwellers in Western inner city areas, but the main drivers were political and came from Washington and London. The key effect was to limit the granting of direct aid and to make all loans (as for the dams) dependent on political and economic 'conditionality'. The result was a weakening of many developing countries' economies, so that the end of the Cold War and the withdrawal of bilateral Soviet or American aid often led to their total collapse rather than the hoped-for transition to democracy. Rather than free governments and open economies, many developing countries have been left with crony capitalism, corrupt leaders, increased poverty and conflict.

More recent thinking about the links between conflict, development and the need to help struggling developing economies

A lecture given by Jeffrey Sachs summed up what might be called the most recent version of the 'classical' view on the links between aid, development and conflict:

> Why does Africa lag? Here is where the scientific evidence on extreme poverty is vital. The overwhelming non-scientific assumption held in our societies is that Africa suffers mainly from the corruption and misman-agement of its leaders. With the viciousness and despotism of Robert Mugabe in Zimbabwe, it's an understandable view. Yet this seemingly self-evident view is wrong as a generalization. Zimbabwe may get the headlines, but there are many countries in Africa, like Tanzania and

Mozambique just nearby, that have talented and freely elected governments struggling against poverty. But they too face great obstacles, and
their people too continue to suffer from extreme deprivation.

(Sachs 2007)

The logic this has inspired is that deprivation must be tackled by
external intervention. Some benign examples have been the provision
of educational programmes to provide all children in developing
countries with access to a laptop computer, and NGO activity to
spread knowledge about the likely vectors of the AIDs virus. There
have also been programmes to promote the 'sustainable development'
that Western agencies believe will enable LDCs to grow while
respecting the environment. In war-torn societies such intervention
has most obviously been by direct international assistance.

Do no harm

Writers like Mary B. Anderson believe that in certain circumstances
aid can indeed support peacebuilding activities, but that carelessly
applied aid will encourage damaging resource transfers and in effect
disempower local people from taking control of their own destinies –
summed up as 'do no harm' (Anderson 1999). Projects governed by
the principle have received high-level backing from Western governmental institutions like Britain's DFID. In one of these, concentrating
on 'conflict sensitivity', the logic is summed up as follows:

> Aid is not neutral in the midst of conflict. Aid and how it is adminis-
> tered can cause harm or can strengthen peace capacities in the midst
> of conflicted communities. All aid programmes involve the transfer of
> resources (food, shelter, water, health care, training, etc.) into a
> resource-scarce environment. Where people are in conflict, these
> resources represent power and wealth and they become an element of
> the conflict. Some people attempt to control and use aid resources to
> support their side of the conflict and to weaken the other side. If they
> are successful or if aid staff fail to recognize the impact of their pro-
> gramming decisions, aid can cause harm. However, the transfer of
> resources and the manner in which staff conduct the programmes can
> strengthen local capacities for peace, build on connectors that bring
> communities together, and reduce the divisions and sources of ten-
> sions that can lead to destructive conflict.
>
> (Conflict Sensitivity Forum 2008–12, http://www.conflictsensitivity.
> org/node/103, accessed 20 July 2015)

Anderson's ideas have certainly struck a chord with those who believe that local solutions to local problems are more likely to be workable and effective than ones imposed by outsiders. It would also be true to say that the freemarketeers (like Bauer) have largely won the day on one aspect of aid, in that it is now appreciated that throwing money into economies that are incapable of using it in productive ways is not very sensible. But it is also appreciated that aid that goes into countries deemed to have a track record of sound market-orientated policies is able to deliver effective, sustained policy reforms. Of course this illustrates how we can see aid as another support for the overarching ideology of the liberal peace.

It also has been widely decided by Western policymakers and IGOs that, as we stated at the beginning of this chapter and earlier in the book, the experiences of the 1990s (especially the abortive intervention in Somalia) and the subsequent events after '9/11' show that development is now a security issue. There was 'a mounting perception and articulation that underdevelopment was dangerous and that – by implication – arising the level of development would increase security in the country and ultimately globally' (Schnabel and Carment 2004, Volume 2).

Aid and 'human security'

Hence much aid in the 1990s and since started to be directed towards DDR programmes with the aim of increasing 'human security', an ideal of seeing development and security as closely inter-linked, which is an unexceptional statement but which needs much greater clarity to operationalize (Cooper et al. 2010; Howe 2013). The concept is often explicitly linked to conflict prevention, with the idea that if the population's security needs can be provided for, the rest of their economic and social existence will be ensured. Aid and conflict prevention measures in general should therefore 'focus on long term, structural challenges to build safe, just and stable societies' (Schnabel in Schnabel and Carment 2004: vol. 2, 109–31). One criticism that can be levelled against the concept of human security is that it also seems to have an inherent belief that basic 'human needs' can indeed be identified, a claim that we have seen in relation to conflict resolution techniques and proposals in Chapter 4. Other commentators on human security are not sanguine as to its usefulness. Marriage comments that it could be seen as so 'infinitely elastic' a concept as to be 'analytically unhelpful. None the less – or maybe because of this – its popularity in policy-making circles has been more enduring' (Duffield 2001; Marriage 2008: 4 and 6; Cooper et al. 2010).

The organization of aid

Partly inspired by this new policy paradigm, since the end of the Cold War a 'humanitarian network' has emerged to deal with both the results of conflicts and aid and reconstruction efforts alike. Since the start of the 'War on Terror' in 2001, conflicts in Afghanistan, Iraq and elsewhere also show that the delivery of aid for purely humanitarian reasons has become in effect even more subordinate to geopolitics than it was during the Cold War. Marriage comments that 'there is abundant evidence that the motivations of donors are mixed and that aid and other interventions are often tempered with blindness or misunderstanding' (Marriage 2008: 2). What, for example, has been the net result of the huge amount of aid that has been, and continues to be, pumped into the Palestinian territories for the local population? How can it be effective when the macro-political situation (the failed Oslo Accords, the Intifada and the brutal simmering war between Israel, Hamas and Hezbollah) means that the aid funds that were intended to bolster the peace process are often of limited effect. Chris Wake paraphrases President Clinton on this case: 'no peace deal can be sustainable if it does not genuinely respond to the needs of everyday people' (Wake 2008: 109–30).

The actors in the delivery of aid are also those present in the full-blown reconstruction efforts discussed in Chapter 5. Donors and deliverers of aid can be bi- or multi-lateral, most usually charitable and sometimes denominational. Humanitarian aid can be delivered by both NGO and IGO organizations as well as, most notably, by the International Committee of the Red Cross (ICRC), which also aims to provide assistance and protection of civilians through the long-established Geneva Conventions. The UN coordinates its activities through the United Nations Office for the Coordination of Humanitarian Affairs (OCHA http://ochaonline.un.org/).

Hoffmann and Weiss (2006) refer to this as a 'humanitarian network', which encompasses 'a constellation of national, IGO, NGO and private contractors that surround a country in crisis, with its own government, local NGOs and victims/recipients of aid' (Hoffmann and Weiss 2006: 122).

The operation of this network obviously varies from case to case. In some areas there will be a great deal of NGO activity, in others very little, as when there is a very poor security situation. To take a particular case, that of Sierra Leone, here all the above actors were (and are) present in a conflict which is both regional, in that it affects the whole of West Africa, but also with specific elements due to the nature of the country itself. The (maybe only provisional) ending of the civil war was particularly engineered by the UK's armed forces and the post-war situation by the UK's development agencies and aid disbursers, although the United States has funded and mainly runs the UN Special Court for Sierra Leone, seen as a major part of the post-war reconciliation process (see Chapter 4). NGOs in Sierra Leone and in Liberia have made great strides in bringing 'human security' to the populations of these two states. No one would dispute the need to get rid of the appalling regimes of Charles Taylor in Liberia and Foday Sankoh in Sierra Leone (Atkinson 2008). But surely it would be prudent to say that the installation of democratic governments in these two places as a result of the expenditure of huge amounts of money by Western states has not really addressed the underlying problems of these countries, which are not so much about democracy as about economic and social hardship and inequality?

One sure conclusion that has to be drawn from the present stage of capitalist development is that where there is no viable state structure the vacuum will tend to be filled by external and internal groups, usually armed to the teeth, and by the immense monetary capabilities of transnational corporations, who will do practically anything to ensure access to their raw material needs. The most obvious example of this is the trade in 'blood' or 'conflict diamonds', a subject that has spawned Hollywood films but has yet to produce any obviously effective codes of conduct for the extraction and end use of such products (see Box 6.1). The Kimberley Process, which was meant to certify the origin of diamonds and prevent conflict diamonds from entering the market, is widely seen as a failure. Diamonds that often grace the necks and hands of the most beautiful women in the West, also lead to the hacking off of equally beautiful necks and hands of women in parts of Africa.

Box 6.1

Blood diamonds and Sierra Leone

The state of Sierra Leone, well endowed with natural resources, including diamonds, but with a succession of weak and corrupt governments, imploded with the end of the Cold War and the emergence of warlords like Foday Sankoh (and Charles Taylor in nearby Liberia) who used terror to extract and monopolize the exploitation of 'blood' diamonds. The international community recognized that markets in the West made such terror profitable and instigated the 'Kimberley Process' to identify and control such processes. This can be seen as being a useful and 'positive start… but only a start' (Grant and Taylor 2004: 399). The key variable has to be seen as the need for a regional, or even global, and not purely national, solution to the dissemination of such materials. The links between the illegal and violent export of such goods has even been explicitly linked into the funding of globally focused groups like Al-Qaeda (Grant and Taylor 2004: 399). Maybe this will be what finally leads to their control, not the suffering of the Sierra Leonean and West African population in general. Sankoh was overthrown by Western-backed mercenaries and British troops, but an uneasy truce is all that can be said to be holding.

Sources: Gberie (2005), Harris (2013), Keen (2005), Richards and Vincent in de Zeeuw (2008).

The problems of aid dispersal

From the 1960s to the 1990s states and international organizations like the World Food Programme (http://www.wfp.org/about), the World Bank and the IMF, became the key distributors of aid, increasingly backed up and after 1990 increasingly supported by NGOs like OXFAM. Official Development Assistance (ODA) was intended to reach 1 per cent of Gross National Product (GNP), a UN target set in the 1980s. However, OECD countries' contributions had, by the end of the 1980s, only increased ODA to about 0.36 per cent of GNP, with the honourable exceptions of states, especially in Scandinavia, and Britain, that have managed to donate up to 0.7 per cent. These sums were often dependent on economic changes, such as the application of 'structural adjustment programmes', and political conditionalities, such as crackdowns on corruption and the implementation of election monitoring. Furthermore, in many cases promises of aid were only partially delivered upon (Forman and Patrick 2000). Marc Williams's comment of 1994 that 'no clear standards exist by which to measure aid effectiveness and to determine under what conditions aid is likely to promote growth' (Williams 1994: 22–3) is still true today. In spite of this, it is often believed that aid will alleviate poverty and help to solve the conundrum outlined earlier by Jeffrey Sachs.

Why is this? First, aid cannot be delivered without security and is often accompanied by corruption. This applies as much to the providers of aid, like the United Nations, as to the receivers. It has been suggested that in effect an 'aid economy' grows up in parallel to a 'war economy', 'where the focus is not so much on benefiting from violence as it is on taking advantage of efforts to relieve suffering' (Hoffmann and Weiss 2006: 107). In some conflicts, such as Bosnia and Afghanistan, maybe in most, aid can fall into the hands of warlords who exploit their position on the ground in an insecure environment to extort 'protection money' from aid agencies (Goodhand, in Barakat 2004). Second, in purely economic terms an obvious unintended result of providing (for example) many tons of free foodstuffs to alleviate a famine situation is to make local production of such foodstuffs uneconomic. Third, aid agencies inevitably create distortions in local employment patterns, as they can offer better salaries to drivers and translators as well as a more general workforce than the local economy can. In some cases, like the Palestinian Authority (PA), virtually the whole economy has become dependent on outside assistance for public sector salaries and, in that case, two-thirds of all government expenditures (Boyce and O'Donnell 2007: 200). Here aid is a positive incentive to corruption, which has arguably had unforeseen political results, including the rise of Hamas as an alternative to the Palestinian Authority in one section of Palestine (the Gaza Strip) and the consequent breaking up of the state (see Box 6.2).

Hoffmann and Weiss argue that recent thinking among Western governments about the role of the private and public sectors (outlined above as a post-Keynesian consensus) has even caused aid to be 'privatized' to some extent, to the point where for-profit organizations have been deployed in an effort to eradicate some of the 'inefficiencies' of the UN and public sector organizations. This has led to some 'murky' deals being struck to deliver security in many developing country conflict zones. For-profit organizations can certainly respond more successfully and quickly, where the UN might have to wait for years in a search for a consensus on a mandate, but of course they risk suffering from a lack of legitimacy as they are seen for what they are, profit-making enterprises (Hoffmann and Weiss 2006: 152–3).

Box 6.2

External assistance and aid to the Palestinian Authority

Part of the peace process that has been under way since the Oslo Accords of 1993 has involved the emergence of an internationally sponsored Palestinian Authority, that it is hoped will one day emerge as part of a 'two-state' solution for Israel and Palestine. The economic assistance given to the PA has been multifaceted and from many sources, including the UN, the EU and a number of individual governments as well as many NGOs. Since 1993 international aid disbursements to Palestinians have totalled around $20.4 billion, averaging $317 per capita annually (Taghdisi-Rad 2011). GDP per capita certainly rose after 1993 under the impact of such aid, but is still low by the standards of the Middle East ($1493 in 2002), and is only up to $2430.7 in 2012 (UNdata 2015). The peace process has stalled and war has returned to Gaza. The key problem is one of a huge dependency on Israel for access to both export markets and imports, a corollary of huge dislike towards Israel, and massive corruption. Some argue that aid has made things worse by damaging local production that is replaced by 'free' donations. 'De-development' has taken place (Tartir 2015). The lack of any discernable improvement of the lot of ordinary people has increasingly pushed them, mainly out of desperation, into violent support for Hamas and other uncompromising political groups, which are often backed by outside powers like Syria (now also plunged into war), Iran and ISIS. The area is a good example of the 'geopolitics of aid' discussed by Keen (2008) and the need for a regional political and economic approach as discussed by Pugh and Cooper (2004).

Sources: Pugh and Cooper (2004), Boyce and O'Donnell (2007): Chapter 7, Keen (2007), Le More (2008), Taghdisi-Rad (2011), Tartir (2015), UNdata (2015).

There is also a clear problem of 'turf wars' developing between IGOs, NGOs and governments. In what David Keen aptly describes as 'complex emergencies' (Keen 2007), having so many players on the field is bound to lead to them duplicating each other, competing for scarce resources, in effect becoming part of the system that is a civil war. One solution that has been suggested, which maybe helps address this problem, harks back to an older model. Rather than country-based policies, we arguably need regional policies, as with the Marshall Plan for Europe, or indeed the UNRRA, the first UN organization that dispersed aid and helped displaced people on global scale in the period 1943–7 (see Williams 2006: 113–22). After all, the 'new' wars do not respect national boundaries any more than 'old' ones did; and the economic issues of the regions affected have much in common (as in Western Africa) (Pugh and Cooper 2004: 80 and *passim*).

A final initial point for consideration might also be that it has to be asked if the weakest are not in effect the least protected? Tales of rape and child abuse have surfaced in a number of UN-run or -sponsored camps in recent years, as far apart as Bosnia and Sierra Leone. The problem has been recognized as acute by the UN, though its internal report was leaked, not published, as of July 2015 (http://www.theguardian.com/world/2015/apr/29/france-promises-act-leaked-un-report-soldiers-child-abuse-claims). In addition 'non-standard' refugees tend to get the least attention from aid disbursement agencies. Keen's research in an earlier Sudanese conflict (in 1984–5) led him to believe that pastoralists (who are a very common population group in much of Africa) and rural dwellers generally get the worst deal in terms of distribution and help, as do more generally disfavoured sections of any society. The telling parallel, he believes, is that of the relief so badly delivered to New Orleans in 2005 (where most people who were neglected were black), or to Thailand and Sri Lanka after the Tsunami disaster of 2004, which was much better delivered to town dwellers and those living in obviously popular tourist zones (Keen 2007: 121–5).

The problems of protecting aid workers

Another result of the 'new wars' and also of the 'War on Terror' has been the breakdown of the idea that 'host' governments can or should protect aid workers to anywhere the same extent that used to be considered normal. A new phenomenon has emerged, that of the 'armed humanitarian', who is seen by some hostile locals as not only delivering aid but also a message of alien 'democracy' and Western mores. For a conservative Afghan elder, bringing education to the women of his area, as UNICEF has done for example, is not a 'neutral' act; it is one that threatens his legitimacy as a lawmaker, and the very culture of his society. More generally we have seen a blurring of the distinction between civil and military actors, even if we must not forget that such distinctions have always to some extent existed. The most obvious example of this blurring is the use of private security companies to defend aid workers, and in many cases to replace regular military forces altogether, as in Iraq (Kinsey 2009). In the case of Afghanistan, Provisional Reconstruction Teams (PRTs) have often mixed civilian and other non-military personnel with soldiers – as Michael McNerney puts it, not so much a 'model' as a 'muddle' (for more on

Plate 15 *Armoured UN vehicles: aid workers have come under increasing threat as humanitarianism has become securitized.*

this see McNerney 2005–6: https://www.rusi.org/downloads/assets/ PRTs_model_or_muddle_Parameters.htm).

As a result, journalists, UN staff and other 'neutrals' have found they are now, like it or not, seen as legitimate targets by warlords. In Afghanistan and Iraq this has particularly been the case, and the UN briefly withdrew its entire staff, humanitarian and otherwise, after the car bomb that killed over 40 UN staff in 2003. They can now only effectively operate from bases inside the security 'green zone' in Baghdad. This may have the advantage of being more secure but also means the UN is less comprehending of the local people and their circumstances (Chandrasekaran 2007). *Médecins Sans Frontières* withdrew from Afghanistan in 2004 for security reasons. Almost one in five humanitarian staff surveyed in 2005 had been a victim of a 'security incident' (Buchanan and Muggah 2005). Many have been killed or kidnapped. Some have suggested that reducing the dissemination of small arms could help solve the problem (Howard 2008: 44), which seems unlikely given the difficulties that such an enterprise would pose in, say, Afghanistan, where possession of a weapon is seen as a sign of basic social status.

Does aid 'work'?

The economic debate since 2000 has started to move against aid as a good idea to help countries both develop and improve their security environments. The word 'trouble' crops up a lot – *The Trouble with Africa* (Calderisi 2006), *The Trouble with Aid* (Glennie 2008), as well as *Culture Troubles* (Chabal and Daloz 2006). Books with increasingly long and complicated titles have proliferated, like William Easterly's *The White Man's Burden: Why the West's efforts to aid the rest have done so much ill and so little good* (Easterly 2006) and Giles Bolton's *Aid and Other Dirty Business: An insider reveals how good intentions have failed the world's poor* (Bolton 2007), and *Africa Doesn't Matter: How the West has failed the poorest continent and what we can do about it* (Bolton 2008).

These latter titles represent an impressive attack from authors who see aid as a form of 'postmodern imperialism', to use Easterly's phrase (Easterly 2006: Chapter 8). All of these writers see military intervention and aid as parts of a liberal peace package that has failed to solve the problems of development and conflict, although not always from the same political perspective.

The basic problem is the one articulated by Roger Riddell: *Does Foreign Aid Really Work?* He finds that aid needs to be drastically 'recast' not eliminated (Riddell 2008: Chapter 15). And the basic solution is summed up in the much remarked upon book by Dambisa Moyo, that aid is somehow *Dead Aid*, that, in the approving words of a preface by Niall Ferguson, it 'encourages corruption and conflict, while at the same time discouraging free enterprise'. So, the logic continues, we need to reduce, or even eliminate aid, which will bring untold benefits to African (and other developing) economies and polities, cutting off the blood supply to the tumours that are poverty and war. Aid, says Moyo, kills off the potential for agricultural countries to develop viable export sectors by dumping huge quantities of 'free' produce on local markets, destroying the incentives for local market development. Worse still, the West's obsession with democracy and good governance, latterly reinforced by visits by rock stars of the 'glamour aid' variety, has had the effect of 'atrophying' the debate on aid (Moyo 2009: 27).

Other critics of aid, like Linda Polman, go even further and assert that aid not only fuels wars, but also prolongs them, an accusation that led to a furious polemic from Bob Geldof, the organizer of the 'Live Aid' concerts which helped deliver aid in the Ethiopian famine of the late 1980s. Other songs released by Geldof since then have received unwelcome criticism along the same lines. One of Geldof's more printable comments was: 'It's a pop song, it's not a doctoral thesis. [Critics] can f*** off' (Polman 2010 and 2011; Geldof 2014).

Conclusions: does aid make conflict worse?

Many of the assumptions about peacebuilding take it as axiomatic that the development of each country takes a similar course, much in line with the classical view of development. Hence, reconstruction that we looked at in Chapter 5 is to install, or reinstall, proper systems of governance and economic development. Even those who believe that such peacebuilding methods are beneficial are aware of the difficulties involved. Roland Paris warns against believing that 'war-shattered states can be hurriedly rehabilitated' (Paris 2004: ix). We might even worry that aid can actually transform power structures and facilitate the emergence of warlord politics. This has been seen as the unhappy experience in Afghanistan in the 1980s and even since the outing of the Taliban and the emergence of the UN-backed governments since 2001 (Goodhand, in Barakat 2004).

As we have seen above, others go much further and blame aid disbursement for the evolution of predatory capitalism itself and for the complicity of the international community in its worse excesses. Though these views (summed up in 'does aid "work"' above) are certainly at the extreme end of the spectrum of criticism of current aid policies pursued by Western governments, IGOs and NGOs alike, the main critique is of the underlying logic of a 'liberal peace' paradigm that only sees what it wants to see, supposedly governments fairly distributing aid to needy people without fear or ideological favour. This is seen as naïve and ultimately self-defeating. Mark Duffield, one major critic of aid, has also accused IGOs and development agencies more broadly of a crude 're-packaging of these aims over the past half-century' that has done nothing to improve the underlying logic or the delivery of aid on the ground (Duffield 2007: 12). The underlying logic for him has been to provide a 'surplus population created through accumulation by dispossession [that] represents life belonging to capitalism'. In this logic what is happening in Darfur, for example, is in the interests of capitalist development but not of the local population.

The critique bears a certain resemblance to that of other writers like Cramer (2006) and Keen (2007), who perceive a variety of errors that derive from the West seeing the problems of development as being those of the aftermath of wars, rather than as due to the problems of a system of internal exploitation of one group by another in most countries that are examined. In effect, slavery has been replaced by a different form of rapacious global capitalism. No solution will, or can, therefore be found without realizing that we need to identify the issues within the political economies of countries like Sudan before there can be any hope of addressing the long-term problems of both development and the conflicts that emerge from this, as the main underlying problems are ones of governance and dominance in developing countries. Neither, as both Cramer and Keen make very clear, can we get far in understanding these conflicts without an appreciation of the importance of violence in development. It is, in Keen's words, a difficult issue of walking the 'tight-rope between explaining and excusing' (Keen 2007: 5). Violence was a great part of the development of Western polities and economies; it is a part of those now undergoing a process of development. Maybe therefore aid can only ever be a sticking plaster on a necessary process of self-harm?

Furthermore, these writers all see the issues involved in development as part of a much wider nexus of liberal governance that perpetuates the North–South dependencies of the colonial period, except that now the main agents of dominance are not governments but Western multinational organizations and IGOs that in effect do their bidding, 'the suppliers, facilitators and cultural sustainers of inhumanity' (Duffield 2007, quoting Slim 1997). The emergence of the 'new wars' has both facilitated and been facilitated by the borderless nature of post-Cold War capitalism, to create what Duffield calls 'network war' (Duffield 2001: 260) or Keen calls 'abusive war systems' (Keen 2007: 9). Aid is for them therefore an essential part of this 'network' or 'system' and must be reformed along with them if it is to help resolve or abate conflict.

The vital questions therefore have to be as to whether aid is only a palliative for much wider problems created by a post-Cold War capitalism that now finds itself unimpeded by Great Power interests and helped by the slow but sure disintegration of the local state structures into a system of local clientelisms that do not respect borders of national loyalties, but only the allure of money. The 'new' wars have created an entrepreneurial class that serves powerful outside capitalist interests (for example in the extraction of diamonds in West

Africa), but is unimpeded by local state authority. In such a situation aid distribution and the organization of refugee assistance becomes just more grist to local warlord interests. It might help assuage the consciences of Western governments and their populations who want to 'do something' to help the starving peoples of the South, but in effect it does the opposite. The botched intervention in Somalia in the early 1990s may be said to be a paradigmatic example of that problem.

May we therefore go back to the question that is asked from opposite perspectives by the conservative views of Luttwak, and his urging to 'give war a chance' (Luttwak 1999), and the much more radical views of Cramer and Duffield, or the more moderate views of Keen, Hoffmann and Weiss? Is aid making the conflicts of Africa and elsewhere worse? Would it not maybe be better not to distribute it at all? Is the problem not the one it has always been – outside interference?

Conversely to follow such advice absolutely would be a counsel of despair, as countless more people would die as a result. In spite of their shortcomings, the international community has defined 'Sustainable Development Goals' to abolish poverty (or even, number 1: 'End poverty in all its forms everywhere'). The Group of Seven (G7, formerly G8 until Russia was suspended in March 2014 over its involvement in the Ukraine crisis) has also made fulsome pledges to reduce poverty. Maybe most important of all, there is a growing awareness among ordinary people in the West that the problems of the developing countries are their problems and that they are willing to support political decisions to help their less fortunate sisters and brothers. More controversially, new donors are emerging, and especially in the Middle East, China and South Korea (Chaturvedi 2012; Mawdsley 2012) who now hold many of the dollar surpluses as a result of globalized trade patterns and other factors. The consensus so far seems to be that these donors are making many of the same errors as the 'old' ones. But that is for the next edition of this book to explore, by which time it is safe to say this will be a burning issue. We have to hope that the debate outlined in this chapter will throw up some better ideas for the aid debaters of the future.

Summary

- Aid – its delivery and logic – has become a key element in understanding contemporary conflicts, and their potential transformation, in the developing world.

- The discussion of aid elicits very different reactions from analysts of development, often depending on the cases they have studied, but also as a result of their intellectual and policy approach.
- Aid is not 'neutral', either in its political logic or distribution.
- The period since the end of the Cold War has seen aid becoming 'securitized'.
- The delivery of aid often requires making difficult moral and political compromises.

Discussion questions

- Can a consideration of the motives for and the delivery of aid help us in an understanding of the dynamics of contemporary conflicts and wars in the developing world?
- What moral and practical problems are there in the delivery of aid?
- Does aid help in peacebuilding?
- Why have aid workers now often become targets in developing country conflicts?
- Should we 'give war a chance'?

Further reading

A comprehensive collection that covers important aspects of the discussion about aid is R. Mac Ginty and J. Petersen (eds) (2015) *The Routledge Companion to Humanitarian Action*, London: Routledge. See also J. K. Boyce and M. O'Donnell (2007) *Peace and the Public Purse: Economic policies for postwar statebuilding*, Boulder, CO: Lynne Rienner. For a general critique of aid see M. Duffield (2001) *Global Governance and the New Wars: The merging of development and security*, London: Zed; M. Duffield (2007) *Development, Security and Unending War: Governing the world of peoples*, Cambridge: Polity; D. Keen (2008) *Complex Emergencies*, Cambridge: Polity; and W. Easterly (2006) *The White Man's Burden: Why the West's efforts to aid the rest have done so much ill and so little good*, London: Penguin. See also (the not uncontroversial) L. Polman (2010) *The Crisis Caravan: What's wrong with humanitarian aid?*, New York: Metropolitan Books; L. Polman (2011) *War Games: The story of aid and war in modern times*, London: Viking.

For a critique of why the international community has gone wrong in its economic and aid policies towards Africa in particular, see P. Chabal and J.-P. Daloz (1999) *Africa Works: Disorder as a political instrument*, Oxford: James Currey; P. Chabal and J.-P. Daloz (2006) *Culture Troubles: Politics and the interpretation of meaning*, London: Hurst; C. Cramer (2006) *Civil War is not a Stupid Thing: Accounting for violence in developing countries*, London: Hurst; and R. Calderisi (2006) *The Trouble with Africa: Why foreign aid isn't working*, London: Macmillan.

Useful websites

(All accessed as at 22 July 2015.) On 'do no harm' and institutions that have signed up for the principle (this site includes many links to other organizations engaged in thinking about or delivering aid in conflict zones), see: http://www.conflictsensitivity.org. Also see (online and in print): OECD (2010) *Do No Harm: International support for state-building*, Paris: OECD. A good summary of the current disasters and responses to them can be found at 'Relief Web': http://reliefweb.int. 'Human security' is best discussed at http://www.humansecurityinitia-tive.org/definition-human-security. The United Nations Office for the Coordination of Humanitarian Affairs (OCHA) can be found at http://ochaonline.un.org/, and the World Food Programme at http://www.wfp.org/. Some useful websites on agencies that disburse aid: http://www.usaid.gov/locations/ [USAID]; http://eur-lex.europa.eu/legal-content/EN/TXT/?uri=URISERV:dv0015 [EU]; https://www.gov.uk/government/organisations/department-for-international-development [DFID, British government agency]; http://www.oxfam.org.uk [OXFAM].

References

Anderson, M. (1999) *Do No Harm: How aid can support peace – or war*. Boulder, CO: Lynne Rienner.
Atkinson, P. (2008) Liberal interventionism in Liberia: towards a tentatively just approach? *Conflict, Security and Development* 8(1): 15–45.
Baldwin, D. (1985) *Economic Statecraft*. Princeton, NJ: Princeton University Press.
Barakat, S. (ed.) (2004) *Reconstructing War-Torn Societies: Afghanistan*. London: Palgrave Macmillan.
Bauer, P. (1991) *The Development Frontier: Essays in Applied Economics*. Brighton: Harvester Wheatsheaf.
Bieri, F. (2010) *From Blood Diamonds to the Kimberley Process: How NGOs cleaned up the global diamond industry*. Farnham: Ashgate.

Bolton, G. (2007) *Aid and Other Dirty Business: An insider reveals how good intentions have failed the world's poor*. London: Ebury Press.

Bolton, G. (2008) *Africa Doesn't Matter: How the West has failed the poorest continent and what we can do about it*. London: Arcade Books.

Boyce, J.K. and O'Donnell, M. (2007) *Peace and the Public Purse: Economic policies for postwar statebuilding*. Boulder, CO: Lynne Rienner.

Buchanan, C. and Muggah, R. (2005) *No Relief: Surveying the effects of gun violence on humanitarian and development personnel*. Centre for Humanitarian Dialogue and Small Arms Survey, UNDP. Available at: www.undp.org/bcpr/smallarms/index.htm (accessed on 8 September 2015).

Calderisi, R. (2006) *The Trouble with Africa: Why foreign aid isn't working*. London: Macmillan.

Chabal, P. and Daloz, J.-P. (2006) *Culture Troubles: Politics and the interpretation of meaning*. London: Hurst.

Chandrasekaran, R. (2007) *Imperial Life in the Emerald City: Inside Iraq's Green Zone*. London: Bloomsbury.

Chaturvedi, S. (2012) *Development Cooperation and Emerging Powers: New partners or old patterns*. London: Zed.

Collier, P. (2007) *The Bottom Billion: Why the poorest countries are failing and what can be done about it*. Oxford: Oxford University Press.

Collier, P. (2008) The bottom billion. Available at: http://www.ted.com/talks/paul_collier_shares_4_ways_to_help_the_bottom_billion.html (accessed on 5 September 2015).

Cooper, N., Pugh, N., Turner, M. (2010) Has Human Security Had Its Day? In Chandler, D. and Hynek, N. (eds) *Critical Perspectives on Human Security: Rethinking emancipation and power in international relations*. London: Routledge, pp. 83–96.

Cramer, C. (2006) *Civil War is not a Stupid Thing: Accounting for violence in developing countries*. London: Hurst.

De Zeeuw, J. (ed.) (2008) *Transforming Rebel Movements After Civil War*. Boulder, CO: Lynne Rienner.

Diamond, J. (2005) *Collapse: How societies choose to fail or survive*. London: Penguin.

Duffield, M. (2001) *Global Governance and the New Wars: The merging of development and security*. London: Zed.

Duffield, M. (2007) *Development, Security and Unending War: Governing the world of peoples*. Cambridge: Polity.

Easterly, W. (2006) *The White Man's Burden: Why the West's efforts to aid the rest have done so much ill and so little good*. London: Penguin.

Forman, S. and Patrick, S. (eds) (2000) *Good Intentions: Pledges of aid for postconflict recovery*. Boulder, CO: Lynne Rienner.

Fukuyama, F. (2005) *State Building: Governance and world order in the twenty-first century*. London: Profile.

Gberie, L. (2005) *A Dirty War in West Africa: The R.U.F. and the destruction of Sierra Leone*. London: Hurst.

Geldof, B. (2014) Bob Geldof tells Band Aid critics to 'F*** off!' *Independent*, 9 December. Available at http://www.independent.co.uk/arts-entertainment/music/news/bob-geldof-hits-back-at-band-aid-critics-they-can-fck-off-9912594.html (accessed on 7 July 2015).

Ghani, A. and Lockhart, C. (2008) *Fixing Failed States: A framework for rebuilding a fractured world*. Oxford: Oxford University Press.

Glennie, J. (2008) *The Trouble With Aid: Why less could mean more for Africa*. London: Zed.

Grant, J.A. and Taylor, I. (2004) Global governance and conflict diamonds: the Kimberley Process and the quest for clean gems. *Round Table: The Commonwealth Journal of International Affairs* 93(375): 385–401.

Guardian (2002) Lord Bauer. Available at: http://www.theguardian.com/news/2002/may/06/guardianobituaries.obituaries (accessed on 10 December 2015).

Harris, D. (2013) *Sierra Leone: A Political History*. London: Hurst.

Hoffmann, P.J. and Weiss, T.G. (2006) *Sword and Salve: Confronting new wars and humanitarian crises*. Lanham, MD: Rowman and Littlefield.

Hopkins, A.G. (2002) *Globalization in World History*. London: Pimlico.

Howard, A. (2008) Small arms vs humanitarian aid: opportunities for action. Available at: www.aidandtrade.org/review-2007-2008 (accessed on 18 July 2008).

Howe, B. (2013) *The Protection and Promotion of Human Security in East Asia*. Houndmills: Palgrave Macmillan.

Hülsemeyer, A. (ed.) (2003) *Globalization in the Twenty-first Century: Convergence or divergence?* London: Palgrave.

Hunt, J. (2004) Aid and Development. In Kingsbury, D., Remenyi, J., McKay, J. and Hunt, J. *Key Issues in Development*. Basingstoke: Palgrave Macmillan, pp. 67–90.

Jacoby, T. (2007) Hegemony, modernisation and post-war reconstruction. *Global Society: Journal of Interdisciplinary International Relations* 21(4): 521–37.

Keen, D. (2005) *Conflict and Collusion in Sierra Leone*. London: James Currey.

Keen, D. (2008) *Complex Emergencies*. Cambridge: Polity.

Kinsey C. (2009) *Private Security and the Reconstruction of Iraq*. London: Routledge.

Landes, D. (1998) *The Wealth and Poverty of Nations: Why some are so rich and some are so poor*. London: Abacus.

Lankester, T. (2012) *The Politics and Economics of Britain's Foreign Aid: The Pergau Dam Affair*. London: Routledge.

Le More, A. (2008) *International Assistance to the Palestinians after Oslo: Political guilt, wasted money*. London: Routledge.

Luttwak, E. (1999) Give war a chance. *Foreign Affairs* 78(4).

Marriage, Z. (2008) Ambiguous agreements: aid in negotiating processes. *Conflict Security and Development* 8(1): 1–13.

Mawdsley, E. (2012) *From Recipients to Donors: Emerging powers and the changing development landscape*. London: Zed.

McNerney, M. (2005–6) Stabilization and reconstruction in Afghanistan: are PRTs a model or a muddle? *Parameters* Winter: 32–6.

Moyo, D. (2009) *Dead Aid: Why aid is not working and how there is another way for Africa*. London: Allen Lane.

Nordstrom, C. (2004) *Shadows of War: Violence, power, and international profiteering in the twenty-first century*. Berkeley, CA: University of California Press.

OECD (2010) *Do No Harm: International support for statebuilding*. Paris: OECD.

Paris, R. (2004) *At War's End: Building peace after civil conflict*. Cambridge: Cambridge University Press.

Polman, L. (2010) *The Crisis Caravan: What's wrong with humanitarian aid?* New York: Metropolitan Books.

Polman, L. (2011) *War Games: The story of aid and war in modern times*. London: Viking.

Pugh, M. and Cooper, N. with Goodhand, J. (2004) *War Economies in a Regional Context: Challenges of transformation.* Boulder, CO: Lynne Rienner.

Riddell, R.C. (2008) *Does Foreign Aid Really Work?* Oxford: Oxford University Press.

Sachs, J. (2005) *The End of Poverty: Economic possibilities for our time.* London: Penguin.

Sachs, J. (2007) Economic solidarity for a crowded planet. Reith Lectures 2007: www.bbc.co.uk/radio4/reith2007/lecture4.shtml (accessed on 27 July 2015).

Schnabel, A. and Carment, D. (eds) (2004) *Conflict Prevention for Rhetoric to Reality: Volume 1: Organizations and Institutions; Volume 2: Opportunities and Innovations; Volume 3: Pacific Settlement of International Disputes.* Lanham, MD: Lexington.

Seitz, T. (2012) *The Evolving Role of Nation-Building in US Foreign Policy: Lessons learned, lessons lost.* Manchester: Manchester University Press.

Slim, H. (1997) *Doing the Right Thing: Relief agencies, moral dilemmas and moral responsibility in political emergencies and war.* Uppsala: Nordic Africa Institute.

Taghdisi-Rad, S. (2011) *The Political Economy of Aid in Palestine: Relief from conflict or development delayed?* London: Routledge.

Tartir, A. (2015) Book review: *The Political Economy of Aid in Palestine: Relief from Conflict or Development Delayed?* by Sahar Taghdisi-Rad. Available at: http://blogs.lse.ac.uk/lsereviewofbooks/2012/07/19/book-review-the-political-economy-of-aid-in-palestine-relief-from-conflict-or-development-delayed/ (accessed on 20 July 2015).

Toye, J. (2014) *UNCTAD at 50: A short history.* UNCTAD: Geneva.

UNdata (2015) (Palestine) https://data.un.org/CountryProfile.aspx?crName=State%20of%20Palestine (accessed on 20 July 2015).

Usher, A. (ed.) (1997) *Dams as Aid.* London: Routledge.

Wake, C. (2008) An unaided peace? The consequences of international aid in the Oslo Peace Process. *Conflict Security and Development* 8(1): 109–31.

Williams, A. (2006) *Liberalism and War: The victors and the vanquished.* London: Routledge.

Williams, M. (1994) *International Economic Organisations and the Third World.* New York: Harvester Wheatsheaf.

Zoellner, T. (2006). *The Heartless Stone: A journey through the world of diamonds, deceit and desire.* New York: St Martin's Press.

Conclusion

The second edition of this book, as with its predecessor, is intended to give those who are relatively new to the fields of both conflict and development a better idea about the debates that we consider to be the most important in understanding the relationship between them. We are aware that we cannot and will not have satisfied all those who read it, as the debates we have summarized are neither the only conceivable ones we could have considered nor by any means those that some would have chosen to emphasize. We are also aware that there are many different approaches that we could have taken, from purely relating what the international development organizations (IGOs) have been doing in conflict zones in an uncritical manner, through to a hard-line 'critical' approach which would take as its point of departure the premise that all Western governments and IGOs are primarily concerned with continuing their former imperialist ventures in a post-colonial environment. To be sure, we do rather veer towards the latter position than towards the former, a tendency that is shown by our privileging of certain discourses over others. But we also freely admit that without IGOs there can be no delivery of better policies, without Western governments there cannot be on the whole the resources and organizational abilities made available to deliver those policies, and without some form of self-interest being manifest there will be no incentive to think about what these policies might be. Globalization has indeed produced many curious antipathies, synergies and disjunctures in thinking and action about development. We all need to understand what those relationships are if we are to improve our game in both the analysis of conflict and the processes of development.

So the approach we have taken in this book has been one premised on the idea that there are serious problems with the way that the twin processes of development and the conflict which is so often attendant on development interact with each other. We do not in any way deny that there has been a huge amount of goodwill and effort put in over many decades by development economists, workers in international organizations, and even politicians, at every level of the global system to try and ameliorate the conditions of underdevelopment and war in which so many of the people on this planet exist. It is true that the 'Millennium Development Goals' and their successor the 'Sustainable Development Goals' set by the UN and other agencies can point to some successes. Absolute poverty levels are down worldwide and in some places, like China, dramatically so. There is less violence if that is assessed across the developing world, but it is more concentrated than in 2009 (cf. Chapter 4).

But it is also clear that there is a growing chorus of criticism of many 'neo-liberal' development policies, not just in the developing world. Aid policies in particular are coming under increasing scrutiny for both their effect and for their intent. This is also true of the continuing debate about the aftermath of the interventions and wars in Afghanistan and Iraq. Liberalism itself is more criticized than it was five years ago, both globally and locally, as the rise of populist and nationalist movements in Europe demonstrates. But we also hope to have shown here that the reflexive nature of liberalism is its greatest strength. It is only because liberalism has been able to constantly reinvent itself that it still dominates global thinking about and the praxis of international relations (Williams 2006).

We certainly do not exempt our own profession of conflict analysis from our worries about some clear policy and intellectual failures. Those who study conflict have, until quite recently, tended to ignore the problems of development. They have also tended to be stuck in the study of very small-scale examples and to take a 'tourist' interest in what is often abject horror and suffering. Very few analysts of conflict are, for example, prepared to spend more than a few months studying a conflict, visiting the area and developing a thorough understanding of its dynamisms over many years. Understanding requires empathy, and empathy requires a lot of contact with the real world of conflict (something that risk-averse universities and insurers frown upon). So if this book has a primary purpose, it is to get a wider debate going between these different constituencies so that the delivery of aid, the arrangements of post-war conflict situations and thinking about related issues might be improved upon in future.

A Conclusion should not be the place to revisit in detail the conclusions at the end of each chapter, but rather to draw out some of the overarching concerns that could be useful in directing future research and action. What might be some of these overarching concerns?

The first is that we are very adamant in our feeling that we need to approach the pitfalls and huge advantages presented by a more rigorous use of historical examples in the study of conflict and development. Aficionados of the 2009 edition will notice that we have included more historical and contemporary literature, especially as it relates to the evolving crisis in the Middle East. We have attempted to better establish clear genealogies of how and why certain terms are privileged in the theoretical and policy discourse, as we will never otherwise understand why these terms are seen in such different ways by their imposers and recipients. *Local* understandings of historical discourses play out in a conflict (Chabal and Daloz 2006), not just our *Western* readings of them. The obvious example that we have talked about at some length is that of 'reconstruction' (Chapter 5). The war in Iraq (2003–the present) shows us that the term can easily be linked into both the histories of the Iraqi people and also into that of those doing the 'reconstructing'. That can in turn evoke positive or negative historical connotations in both principal parties of the relationship. The alternative models that can be then generated can either exacerbate an existing conflict within a country or help it come to some kind of successful conclusion. There is already some suggestion that the people of the Middle East are now conceiving and practising more indigenous forms of 'reconstruction' through grass-roots organizations, such as Hamas or Hezbollah, which have local credibility, but which evoke hostile reactions from many governments and IGOs. But if the IGOs and governments' activities give the impression of 'colonialism' to the recipients, can they work? Equally, in trying to 'resolve' a conflict or 'mapping' it, there is no use in the outside world imposing its interpretation of how history should be viewed. Again in line with Chabal and Daloz (2006), we believe the 'interpretation of meaning' has to be local.

On a negative note, we have also tried to show where an understanding of history can actually exacerbate a conflict. In many areas of the world unscrupulous elites have stressed 'their' version of history to denigrate or even to incite their enemies to a hatred they did not previously feel. In the Middle East, for example, we can see that historical misunderstandings have become an embedded part of the discourse of conflict on all sides. But there is certainly a need for more investigation of how this might apply in many African

Plate 16 *Ruined tower blocks in Beirut: Hezbollah are organizing local reconstruction efforts.*

countries, where the historical and anthropological literature needs to be far better understood by those who study conflicts in those areas. We are far from understanding how history 'plays' in many conflicts in developing countries, either as a force for peacebuilding or as one for peace destroying. One thing is clear, however: the conflicts or humanitarian emergencies that we might read about in our news-papers did not begin with the last incident. Instead, we need to transcend the tendency towards ahistoricism and realize that current conflicts are often merely the latest installment in a complex history of conflict, identity changes, population flows, border redrawings and other long-term social, economic and cultural dynamics.

The next major concluding thought is that we are driven to believe that 'bottom–up' activities for peacebuilding are more likely to have positive long-term effects than 'top–down' approaches. The dangers of giving people the impression that they are having something passed down to them, rather than doing something for themselves, is bound to lead to a sense of disempowerment and alienation. If this is then reinforced by the feeling among a local population that the incoming ideas and power are not really there 'for them', then the development or peace that is hoped for will not do much to help solve the conflict in question. The presence of peacekeepers is, for example, often seen as necessary, but cannot be of much use to help a process of peace and development if those peacekeepers distort the local economy, suck out the best labour power, and generally lord it over the local population without contributing much to their security or well-being. The key point is that an appropriate balance between indigenous and external norms and capabilities needs to be struck. Of course, this is easier said than done and the nature of the balance will differ from context to context. But a good starting point for potential interveners is to move beyond a position of assuming that they have all of the answers. Instead, rather than viewing local populations as victims, recipients, dependents or troublemakers, it is important that they are viewed as change-agents with substantial capabilities. We must be careful not to romanticize everything local and bottom–up. But, by and large, local approaches to peacebuilding and development can be sustainable and in tune with on-the-ground cultural norms.

Third, we would not be the only, but we would like to be the latest, couple of thinkers to doubt the all-embracing truth of the 'liberal

Plate 17 *Stop sign in Jordan: care must be taken in the transfer of Western ideas and practices to non-Western contexts.*

peace' thesis. It may well be true, even empirically, that 'democracies do not go to war with one another'. But the point is surely that in the case of all developing countries struggling with, or attempting to emerge from, conflict the basic principle does not pertain? Democracies in Europe took many hundreds of years to hone their conflict-resolution skills, ones that still desert them quite regularly. The countries of Africa and elsewhere have not in many cases ever achieved full 'state-ness', never mind democracy. Maybe they cannot without a long-drawn-out process of violence, as some claim from a variety of standpoints. But even if they can, they surely should be allowed to do it without undue outside interference? Whether it is democracy at gunpoint (as in the cases of Iraq and Afghanistan) or democracy by PowerPoint (as in the case of hectoring speeches by Western leaders when they visit developing countries), if democracy does not take root in villages, towns and cities across a country, then it will not thrive.

As a corollary to this point we believe that many of the examples we have explored here show that the uncritical use of neo-liberal economic policies have done much to exacerbate the bad odour in which many 'good' Western principles are held in the developing world. Democracy is *per se* a system that most aspire to, but the relentless way in which the World Bank and other organizations have insisted on 'conditionality', 'big bang' changes and structural adjustment has done much to damage the view of those at the local level. This has been observed of course by World Bank officials themselves, with Stiglitz (2003) the most prominent among them. Others, like Amartya Sen (2001), have noted that poverty is the greatest enemy of democratic change and have lambasted the IGOs for their narrow emphasis on empowering the elite in developing countries. Importantly, poverty is relatively straightforward to deal with if there is generalized economic growth: everyone – rich and poor – can do well. Inequality, however, is another matter, and some (the rich or relatively rich) will have to give something up if those poorer than them are to benefit. That involves hard choices that rich countries and international organizations, by and large, have avoided.

Conflict theory may have something to say here in a general way to the development economists and governance experts. Conflict resolution theorists (as was stressed in the Introduction), long derided as 'cranks' for advocating the empowerment of the grass roots of societies, are now viewed as having seen the coming of elements of a 'post-state' world (through globalization) well before the realist/statists did.

What once was considered equally crankish, such as the environment or even the study of unconventional warfare, is now seen as pretty mainstream. Where only states were taken as the units of analysis in both conflict and strategic analysis as well as in economics, sub-state actors, structures and individuals are now routinely evoked in suggesting both why problems occur and how to deal with them. IGOs will have to adopt this understanding to a far greater extent if they are to deal with today's and tomorrow's problems in the developing world. We may even have to break open that ultimate realist shibboleth of 'sovereignty' and contemplate reparcelling the states of Africa and elsewhere, or, alternatively, look to much more regional power sharing and identity-based policies. Kikuyu, Hutu, Tutsi or Xhosa people across Africa have to feel they can be given their recognition and dignity without only being referred to as 'Kenyan', 'Rwandan' or 'South African'. This is in much the same way as we have let the local populations of their respective areas sort out their identities in Ireland, the Former Yugoslavia or Former Soviet Union without losing all of the benefits that those 'unitary' bodies once provided. How we keep what is best about old structures, while allowing the emergence of new ones, is not a problem in the developing countries alone.

Fourth, we should be aware that the context in which development and conflict occurs is changing rapidly. This change stems from many sources: an increasing global population and the consequent pressure on resources; the changing nature of power relations as China, India, Russia, Brazil and others stake their claim to be regional and world players; the declining importance of sovereignty in a globalizing world; the ever-present and growing dissatisfaction with the United Nations and other multilateral institutions; or the growing significance of China and some Gulf and Arab states as key players in development and humanitarian interventions. The key point is that many of the lenses that we have used to analyse conflict and development over the past decades will need to be radically reassessed to take account of the changing nature of conflict and development processes. What we think we know now, may not necessarily be fit for analytical or practical purposes in the future.

Crucially, travel to many non-Western contexts reveals an important truth: these societies are thriving, dynamic and bursting with inventiveness. It is not the case that the people of the developing world are somehow sitting around waiting for development to come to them. Many are fashioning their own preferred types of development and conflict avoidance through trade, cultural activities and family and community life. They are grasping opportunity and creating their own modernity that might be inflected with local and traditional ways of doing things. It may be a cliché, but travel to Somalia and you will see camel traders with the latest smartphone. The 'remotest' trading centre in Uganda will have lots of cooking utensils for sale – made in China and transported through complex trading networks. If anyone is going to be 'left behind' it is those in the 'developed world' who do not update their thinking on the nature of the developing world.

Last, but not least, we hope we have demonstrated that the problems of conflict and development are ones that affect all of us, wherever we live, and that there is a unity of the human spirit and of human suffering that makes us all morally responsible for each other's welfare. This is also true when we are attempting to impose democracy promotion and statebuilding in conflict or non-conflict situations. We have endeavored throughout this book to see development as both 'top–down', the work of institutions (Chapter 2), and as something that affects ordinary people (Chapter 3 and elsewhere). That does not mean that we should be prescriptive about the moral universe the other lives in, but it does mean that we have to take responsibility for the negative effects of our actions in the West when we think about the impact they will have in developing countries. We need a new ethic of development as we need a new ethic of international relations. That will have to be for another book in this series to think about.

References

Chabal, P. and Daloz, J.-P. (2006) *Culture Troubles: Politics and the interpretation of meaning*. London: Hurst.

Sen, A. (2001) *Development as Freedom*. Oxford: Oxford University Press.

Stiglitz, J. (2003) *Globalisation and Its Discontents*. London: Penguin.

Williams, A. (2006) *Liberalism and War: The victors and the vanquished*. London: Routledge.

Index